MOLECULAR
BIOLOGY
INTELLIGENCE
UNIT

POLYMERASE CHAIN REACTION (PCR)
THE TECHNIQUE AND ITS APPLICATIONS

Rosalind A. Eeles, M.A., M.B., B.S., M.R.C.P., F.R.C.R.
Alasdair C. Stamps, Ph.D.

The Institute of Cancer Research
The Royal Marsden Hospital
Sutton, Surrey
United Kingdom

R.G. LANDES COMPANY
AUSTIN

MOLECULAR BIOLOGY INTELLIGENCE UNIT

POLYMERASE CHAIN REACTION (PCR): THE TECHNIQUE AND ITS APPLICATIONS

R.G. LANDES COMPANY
Austin

CRC Press is the exclusive worldwide distributor of publications of the Molecular Biology Intelligence Unit.
CRC Press, 2000 Corporate Blvd., NW, Boca Raton, FL 33431. Phone: 407/994-0555.

Submitted: May 1993
Published: August 1993
Second Printing: April 1994
Third Printing: July 1994

Production Manager: Judith Kemper
Copy Editor: Constance Kerkaporta

Please address all inquiries to the Publisher:
R.G. Landes Company, 909 Pine Street, Georgetown, TX 78626
or
P.O. Box 4858, Austin, TX 78765
Phone: 512/ 863 7762, FAX: 512/ 863 0081

ISBN 1-879702-58-4
CATALOG # LN0258

Library of Congress Cataloging-in-Publication Data
Eeles, Rosalind A., 1959-
Polymerase chain reaction (PCR):the technique and its applications /
Rosalind A. Eeles, Alasdair C. Stamps.
p. cm.—(Molecular biology intelligence unit)
Includes bibliographical references and index.
ISBN 1-879702-58-4 : $89.95
1. Polymerase chain reaction. I. Stamps, Alasdair. II. Title. III. Series.
QP606.D46E34 1993
574.87'3282—dc20
93-4961
CIP

CONTENTS

ACKNOWLEDGMENTS

This monograph was written over a period of several months by a Research Scientist (A. Stamps) and a Medical Clinical Scientist (R. Eeles). We hope that readers will find the book useful if they wish to know about the enormous potential of PCR and/or if they want to set up experiments using PCR themselves. We have tried to include lessons that we have learned in using this technique so that others can benefit from our learning curves. We are particularly grateful to Christine Bell for her help with typing the manuscript and to Ron Landes for instigating this project. This monograph is dedicated to our long suffering partners, Ginnie and Doug.

Ros Eeles
Alasdair Stamps
Sutton
April 1993

PREFACE

The polymerase chain reaction (PCR) has revolutionized molecular biology. Before its inception, DNA sequences had to be amplified by time-consuming cloning. This took several days whereas the PCR can be set up and completed in a matter of hours. The PCR is an in vitro method of amplifying DNA sequences exponentially. In keeping with this there has been an exponential explosion of applications of the technique both in clinical and scientific research. This monograph places the PCR in its historical context and describes the basic technique assuming no prior knowledge of molecular techniques or laboratory methods. It discusses the PCR environment and outlines the problems and pitfalls, providing tips for overcoming them. The variations of the technique and scientific and medical applications are discussed. We hope this monograph will be of interest to a varied audience from those with no previous laboratory experience who are interested in setting up the technique, to those who routinely use PCR and wish to widen its application.

═══CHAPTER 1═══

PCR: A HISTORICAL PERSPECTIVE

The polymerase chain reaction (PCR) created a revolution in molecular biology research and its applications. PCR is an in vitro method that enzymatically amplifies specific DNA sequences using oligonucleotide primers that flank the region of interest in the target DNA. The principle involves a repetitive series of cycles each of which consists of template denaturation, primer annealing and extension of the annealed primers by a DNA polymerase to create the exponential accumulation of a specific fragment whose ends are determined by the 5^1 ends of the primers. The PCR is so named because it involves a polymerase and the products synthesized in each cycle can serve as templates in the next so the number of DNA copies approximately doubles at every cycle to create a chain reaction similar to the principles in a nuclear reactor. In practice the amplification is not a hundred percent efficient and in 20 cycles will be about 10^6 to 10^8 fold. The reaction is therefore equivalent to finding one specific person in San Francisco and cloning them so that they could then populate the whole city. The concept of the polymerase chain reaction was first devised by Kary Mullis[1,2] and was applied by a group in the Human Genetics Department at Cetus Laboratory to the amplification of human betaglobin DNA.[3]

The method was revolutionized by the introduction of automatic thermal cyclers and a thermostable DNA polymerase (Taq polymerase). Initially PCR used the Klenow fragment of *E.coli* DNA polymerase 1 but this enzyme is inactivated by the high temperature needed for DNA denaturation at the beginning of each cycle, so fresh enzyme had to be added during each cycle. The different temperatures for each part of the cycle were obtained by moving the reactions to different water baths at the relevant temperatures.

The introduction of a thermostable DNA polymerase, Taq polymerase, first isolated from *Thermus aquaticus* enabled the amplification reaction to be carried out by cycling the temperature within the reaction tube after mixing all the reaction components.[4] *T. aquaticus* strain Y T 1 is a thermophilic bacterium capable of growth at 70-75°C. It was originally isolated from a hot spring in Yellow Stone National Park and described first in 1969.[5] The

Polymerase Chain Reaction (PCR): The Technique and Its Applications, written by Rosalind A. Eeles, M.A., M.B., B.S., M.R.C.P., F.R.C.R.; Alasdair C. Stamps, Ph.D.; © 1993 R.G. Landes Company.

optimum temperature for DNA synthesis is between 75 and 80° C and the enzyme can incorporate about 150 nucleotides per second per enzyme molecule. The introduction of Taq polymerase not only simplifies the procedure of PCR but also increases the efficiency and product yield because the higher optimum temperature allows the use of higher temperatures for primer annealing and extension, thereby increasing the overall stringency of the reaction. At 37° C which is the optimum temperature for the Klenow enzyme, non specific annealing can occur, resulting in nonspecific amplification products. Taq polymerase also enables the amplification of much longer fragments (up to 10Kb) although these are amplified with reduced efficiency.[6] The Klenow enzyme only enabled amplification up to about 400 base pairs.

In the last few years, the important advances in PCR have been modifications of the technique to enable amplification of DNA from archival material[7] making large data banks of paraffin embedded material available for study; modifications of the technique also enabled PCR to be performed from single cells exploiting the techniques unique capability of massive amplification[8]. Extension of automation using robotics has enabled large numbers (about 100) of PCRs to be done at one time and the total set up time can be reduced to about 15 minutes (Biomek™, Beckman, USA). Advances in enzyme technology have introduced new thermostable DNA polymerases which have proofreading ability which Taq polymerase lacks, reducing the incidence of polymerase-induced errors in nucleotide incorporation. The importance of the PCR technique is reflected by the exponential increase in the number of publications relating to PCR from 3 in 1986 to a staggering 1700 in 1990. Many reviews and several textbooks have been written and examples are listed in the references.[9-17]

REFERENCES

1. Mullis KB, Faloona F. Specific Synthesis of DNA in vitro via a polymerase catalysed chain reaction Meth Enzymol 1987; 155:335-350.
2. Mullis KB, Faloona F, Scharf SJ, Saiki RK, Horn GT, Erlich HA. Specific enzymatic amplification of DNA in vitro: The polymerase chain reaction—Cold Spring Harbor Symp Quant Biol 1986; 51:263-273.
3. Saiki R, Scharf S, Faloona F, Mullis K, Horn G, Erlich HA, Arnheim N. Enzymatic amplification of b-globin genomic sequences and restriction site analysis for diagnosis of sickle cell anaemia. Science 1985; 230:1350.
4. Saiki RK, Gelfand DH, Stoffel S, Scharf S, Higuchi RH, Horn, GT, Mullis KB, Erlich HA. Primer-directed enzymatic amplification of DNA with a thermo-stable DNA polymerase. Science 1988; 239:487-491.
5. Brock TD and Freeze H: Thermus aquaticus gen.n. and sp.n. a non-C sporulating extreme thermophile. J Bacteriol 1969; 98:289-297.
6. Jeffreys AJ, Wilson V, Neumann R, Keyte J. Amplification of human minisatellites by the PCR: Towards DNA fingerprinting of single cells. Nucleic Acids Res 1988; 16:10953-10971.
7. Impraim CC, Saiki RK, Erlich HA, Teplitz RL. Analysis of DNA extracted from formalin-fixed, paraffin-embedded tissues by enzymatic amplification and hybridization with sequence-specific oligonucleotides. Biochem Biophys Res Commun 1987; 142:710-716.
8. Spann W, Pachmann K, Zabnienska H, Pielmeier A, Emmerich B. In situ amplification of single copy gene segments in individual cells the polymerase chain reaction. Infection 1991; 19; 4:242-4.
9. Remick DG, Kunkel SL, Holbrook EA, Hanson CA. Theory and applications of the polymerase chain reaction. Am J Clin Path 1990; 93:Supp 1, S49-S54.

10. Rodu, B. The polymerase chain reaction: The revolution within. Am J Med Sci 1990; 299:210-216.

11. Wright PA, Wynford-Thomas, D. The polymerase chain reaction: miracle or mirage? A critical review of its uses and limitations in diagnosis and research. J Pathol 1990; 162:99-117.

12. McCormick, F. The polymerase chain reaction and cancer diagnosis. Cancer Cells, 1989; 1:56-61.

13. Eeles RA, Warren W, Stamps A, The PCR Revolution, Eur J Cancer 1992; 28:1 2289-2293.

14. PCR Technology. Principles and applications for DNA amplification by Erlich HA, Stockton Press 1989.

15. Current communications in molecular biology: The polymerase chain reaction. Editors Erlich HA, Gibbs R, Kazazian HJr, Cold Spring Harbour Laboratory Press 1989.

16. PCR protocols. A guide to methods and applications. Innis MA, Gelfand DH, Sninsky JJ,White TJ (eds.) 1990:Academic Press, Inc.

17. Genome Analysis: A practical approach. Davies KE, ed. IRL Press 1988; 141-152.

========CHAPTER 2========

THE POLYMERASE CHAIN REACTION

As with any new technology, the decision to apply PCR, how to apply it and in which way to adapt it to suit the task in hand depend on a working knowledge of the system. This and the following chapter have been included to address the basic theory of PCR and discuss the practical limitations associated with its application.

As with other molecular biological methods such as DNA sequencing and labelling, the polymerase chain reaction depends on the activity of a natural DNA synthesizing enzyme. Instead of simply making a single complementary 'copy' of any DNA present in the reaction vessel, however, PCR repeats this process many times over, amplifying the DNA on a logarithmic scale, and is moreover capable of replicating a specific DNA sequence out of a mixture of many millions of different sequences. Thus the technological advance of this method is twofold: amplification of infinitesimally small samples is now possible and this process can be made specific to one sequence.

THE POLYMERASE CHAIN REACTION CYCLE

All PCR applications include the three main steps: denaturation, annealing and polymerization, which are shown diagrammatically in Fig. 1. These three steps make up the PCR cycle.

DENATURATION

The DNA sequence which is to be amplified by PCR is known as the template. It is usually a fragment of double-stranded DNA. Taq polymerase,[1] the DNA synthesizing enzyme most often used in PCR, requires DNA templates to be single-stranded before it is able to make a copy. The template must therefore be denatured into two complementary single strands of DNA before the reaction can commence. This is most easily accomplished by briefly heating and rapidly cooling the DNA, an action which would

Polymerase Chain Reaction (PCR): The Technique and Its Applications, written by Rosalind A. Eeles, M.A., M.B., B.S., M.R.C.P., F.R.C.R.; Alasdair C. Stamps, Ph.D.; © 1993 R.G. Landes Company.

destroy the activity of most enzymes, but only slightly affects the activity of the heat-stable Taq polymerase.

ANNEALING

In common with almost all DNA synthesizing enzymes, Taq polymerase joins deoxyribonucleotides onto the 3' end of one DNA strand, called the primer, and is only able to carry out this activity when the primer is annealed in complementary fashion to the DNA template strand. In a population of denatured DNA strands, synthesis of new DNA can be made highly specific by supplying an excess of defined, short DNA sequences of 17 to 70 nucleotides, called oligonucleotides or "oligos". Oligos can be chemically synthesized in sequences which are complementary to specific sequences of interest. After denaturation, the reaction is quickly cooled, preventing immediate reannealing of long DNA strands. Due to their small size, oligos now rapidly anneal to the single strands of DNA at positions containing the specified template sequence. In these positions, they act as primers for Taq polymerase. The terms "oligo" and "primer" are thus used interchangeably in this context. In practice, formation of the specific primer-template complex is highly temperature dependent, so for annealing to take place, the temperature of the reaction must be lowered to a preset level calculated to maximize primer-template interaction.

POLYMERIZATION

Also known as extension, this is the final step of the PCR cycle in which the temperature of the reaction is adjusted to the optimum for Taq polymerase activity. During this step, the polymerase enzyme incorporates nucleotides into the nascent DNA strand, producing a complementary copy of the DNA template in the region specified by the annealed primer. The new temperature is invariably above that at which annealing occurs,

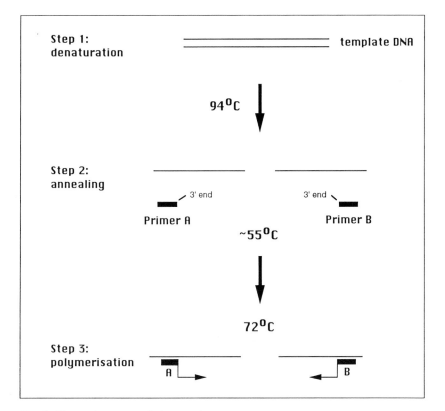

Fig. 1. The polymerase chain reaction.

but does not lead to denaturation of the primer-template complex, presumably because the enzyme is already active at the annealing temperature and significantly increases primer length during the annealing step, thus raising its denaturation temperature above that of the polymerization step.

THE MECHANISM OF AMPLIFICATION

The three steps of the cycle described above are used to synthesise a complementary strand of DNA in the same way as, for instance, primer extension, with annealing temperature variations being used to confer specificity upon primer-template complex formation. The polymerase chain reaction, however, differs from other DNA synthesizing reactions in that a defined sequence is copied repeatedly in a manner which produces logarithmic amplification. This is achieved using two different primers which are each complementary to opposing strands of the template DNA, at a known sequence distance apart on that template (Fig. 2). The PCR cycle is then repeated many times to amplify the template, as follows.

1ST CYCLE

In the first polymerization step a new DNA strand is synthesized onto each primer, giving a nucleotide sequence complementary to the original template. The polymerase enzyme will continue to extend this new strand for the duration of the polymerization step. This means that new sequences primed from oligo A, for example (Fig. 2), will extend past the

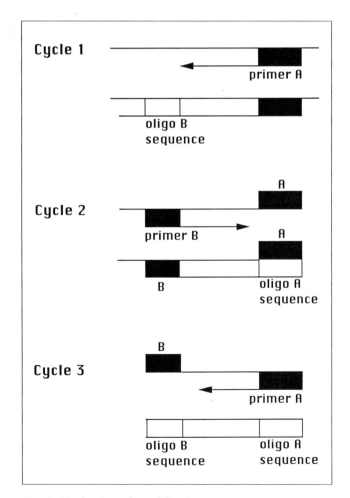

Fig. 2. Mechanism of amplification.

oligo B sequence on this template strand, producing a new DNA strand which starts with the oligo A sequence and contains sequences complementary to oligo B. The same occurs on the other template strand, except that primer B forms the starting sequence. The original template has thus effectively been copied in this region, producing two DNA molecules which may be copied again in the next cycle.

2ND CYCLE

The DNA is again denatured in the first step of the second cycle. More primers anneal in the second step, both to the original template strands and to the new DNA strands produced in the first cycle. Polymerization then proceeds as before, but in the latter case, synthesis of new DNA terminates when it reaches the 5' end of the template, represented by the primer of the first cycle. Thus, if oligo B primes new DNA synthesis on the oligo A-primed strand of cycle 1, this new complementary DNA strand terminates at the 5' end of oligo A (Fig. 2) as there is no further template DNA beyond this point. The length of the new strand produced in this way corresponds to the distance between the two PCR primers.

3RD CYCLE

The DNA strands of defined length synthesized in the second cycle are reprimed with the opposite primer in the third cycle. Polymerization leads to the synthesis of two copies of double-stranded DNA of the defined fragment size.

The number of these fragments is doubled during each PCR cycle, while the original template strands are copied only once. The double-stranded DNA copies therefore become the major reaction product, accumulating logarithmically.

If copying efficiency were 100%, a 30-cycle PCR would amplify a given template by 2^{30}. Thus a single DNA template could potentially be amplified to $\sim 10^9$ molecules. For a typical PCR DNA fragment of 250 base pairs, this is approximately 0.3 ng of DNA, easily detectable by Southern blot hybridization. Twenty such DNA templates similarly amplified would yield enough PCR products to produce a visible band in an ethidium bromide stained electrophoresis gel.

COMPONENTS OF THE REACTION

The Taq polymerase enzyme works with high efficiency in relatively simple buffer systems. Thus the essential components of the polymerase chain reaction include: the DNA template, two defined oligos to act as primers, the four deoxynucleotide triphosphates (dNTPs) for incorporation into the new DNA strands, Taq polymerase and magnesium, which is the cationic cofactor of the enzyme, in a Tris buffer of appropriate pH and salt concentration. The salt most frequently used is potassium chloride. A layer of light mineral oil is always placed on top of each reaction to prevent evaporation which would otherwise stop the reaction within a few cycles.

These basic components remain largely unchanged for most applications of PCR, but their concentrations and the buffering pH of the reaction vary according to the type of experiment, the nature of the primers and/or template and the experience of the experimenter. Many of the early publications on PCR describe very varied reaction buffers and starting concentrations of primers and dNTPs.[2-8] This may have been mainly due to the recent discovery of Taq polymerase as the PCR enzyme, and reflects the fact that a reasonable level of amplification can be achieved even with crude preparations of Taq under suboptimal reaction conditions, given a high enough input of template DNA.

The reaction mix is therefore described in Table 1 in terms of lower and upper limits of pH or concentration, and choice of detergent or protein buffer.

The authors have tested a number of combinations of these components and concentrations. For the more straightforward reactions in which there is a good supply of purified template DNA, e.g., 10 ng of genomic DNA, with average length oligos (around 20 nucleotides), the protocol described in Table 2 will give a good yield of PCR products with high specificity using most commercially available Taq polymerases. Mg^{2+} concentration, which affects primer specificity, is dealt with in Chapter 3, "Managing the Method".

In reality, there is no such thing as a standard PCR mix, as the efficiency and specificity of the technique depends on a number of variables, which are affected to different degrees by, for instance, primer sequence and template source. Thus the reaction must be calibrated for each new primer-template combination. Factors affecting PCR, their calibration and the avoidance of artifacts are described in the next chapter.

THE THERMAL CYCLE

The three steps of the polymerase chain reaction are initiated by temperature changes. The temperatures chosen for each step are crucial to the specificity and efficiency of the chain reaction phenomenon as are the times spent on each step and in adjusting temperatures between steps.

DENATURATION

The reaction mix must initially be heated sufficiently to denature long stretches of double-stranded DNA. For the buffer described in Table 2, DNA undergoes rapid denaturation at 94°C. This is important, as degradation of DNA is accelerated at such elevated temperatures, as is the breakdown of the enzyme. "Heat stable" Taq polymerase, like all enzymes, loses activity at a rate which is proportional to temperature. Thus, enzyme activity is lost at a much higher rate while template DNA is being denatured. A compromise must therefore be found between allowing enough time for the template DNA to denature and reducing the total time spent at 94°C. If, say, five minutes are allowed for each denaturation step in a 30-cycle PCR, the enzyme will be subjected to 94°C for a total time of 2.5 hours. Even though the template will be thoroughly denatured for each cycle, the enzyme may fail halfway through the run. In practice, 30 seconds to 2 minutes are allowed for the denaturation step, with 1 minute being the optimum choice for most templates. Nor is a long initial denaturing step necessarily beneficial, as it may reduce enzyme activity before it even starts working.

Table 1. Concentration and pH ranges of the PCR mix

Reaction component	Range of values	References
Tris-HCl	10-67 mM	
	pH 8.2-9.0	2,4-7
KCl	25-50 mM	2,6,7
Gelatin	0.01-0.1% (w/v)	2,6,8
Triton-X-100	0.01% (v/v)	
Tween 20	0.05% (v/v)	6
BSA	0.01-0.1% (w/v)	2,6
$MgCl_2$	see Chapter 3	
dNTPS (each)	50-200 µM	2,3,6,8
Primers (each)	10-250 pmol/reaction	2,3,6,8
Taq polymerase	1-2.5 U/reaction	2,3,6,8

ANNEALING

Following denaturation, the DNA must be cooled rapidly to prevent reannealing of the original strands, which would either stop primer annealing or obstruct the progress of the enzyme along the template strands. Rapid cooling is achieved with greater or lesser efficiency by the various commercially available thermal cycling machines (described below). The annealing temperature and time may be crucial to the specificity and sensitivity of the reaction. Annealing temperature is dependent on the sequences of the primers and the ionic strength of the buffer system and will be dealt with in Chapter 3. The time spent at the predefined annealing temperature is again somewhat of a trade-off: if left too long, mispriming (annealing of oligos to the wrong sequence) or template reannealing may occur; if too short, insufficient primer annealing may occur, particularly when primer concentrations decline in the later cycles of a long PCR run. Again, 30 seconds to 2 minutes are usually allowed for this step. Shorter times give higher specificity but lower yield.

POLYMERIZATION

Taq polymerase has an optimum operating temperature of 75-80° C[9], although the polymerization step in most PCR protocols is set at around 72-75° C. Although this is higher than the annealing temperature in the vast majority of cases, raising the temperature for

---

Table 2. Standard PCR components

10	mM Tris-HCl pH 8.4
50	mM KCl
1	mM mgCl2
0.05%	(v/v) Tween 20
0.2	mM dNTPs
100	pmol each primer
1	U Taq polymerase

---

polymerization does not cause denaturation of the primer-template complex as it is rarely anywhere near the 'melting temperature' (T_m) of the complex. T_m is the temperature at which 50% of double-stranded nucleic acid species are denatured. For an oligo consisting of 50% guanosine and cytosine, in a typical PCR buffer of 50 mM KCl, T_m is 80.4°C. Therefore primers with this G+C content, once annealed, are unlikely to denature at 72-75°C. In addition, the polymerase is active at the annealing temperature, so it is likely that the primers will be considerably extended by the time the polymerization temperature is reached, further stabilizing the double-stranded complex.

The length of the polymerization step is adjusted to suit the rate of incorporation of nucleotides by Taq polymerase. At the optimum temperature, this is up to 150 nucleotides per second,[9] but in practice polymerization times are longer than might be expected to get a good yield of PCR products. In general, a 1 minute polymerization step is sufficient for PCR products up to 1 kilobase in length; larger fragment sizes require proportionally more time for efficient amplification.

THERMAL CYCLERS

The advent of PCR coincided with the appearance of a number of programmable thermal cycling machines on the laboratory sales market. The price of these has steadily declined due to fierce competition, but more recently companies have tried to produce PCR machines which carry out more and more esoteric tasks. In many laboratories there now tends to be a glut of thermal cyclers as efforts are made to buy new machines which, for instance, handle 96-well microtiter plates, or heat samples more quickly between PCR steps.

Almost all PCR machines are based around a heating block designed to fit 0.5 ml microcentrifuge tubes. The temperature of the block is controlled by a programmable microprocessor. It is always electrically heated, but cooling systems vary. A few are cooled by running water, but this frequently results in block failure due to 'furring up' of the cooling channels by mineral deposits precipitated out by the high temperatures. Most thermal cyclers are air-cooled by an integral fan which switches on when cooling is required. These machines are usually very reliable, but some temperature variation may occur across the block. The most sophisticated machines use electrical Peltier pumps to operate the heating block. These provide the option to control temperature 'ramping', i.e., the rate of cooling, which can increase the efficiency of primer annealing, and have the advantage that the block may be cooled to below ambient temperature. PCR products may therefore be rapidly cooled to 5°C immediately after the last cycle, preventing the degradation of DNA which may be caused by exposure to room temperature overnight after an 'out of hours' PCR run. The disadvantage of these machines is in the initial outlay, and some of the earlier machines proved to be unreliable.

Some PCR machines have electronic accessories which monitor each cycle, providing a record of temperature conditions during each run. Although these devices produce an unnecessary amount of data for each run, they can be useful in troubleshooting if, for instance, a regularly used reaction system suddenly fails. Most good thermal cyclers are supplied with a reference tube, which is a sealed 0.5 ml microcentrifuge tube containing a thermocouple. This gives the option of monitoring temperature within the reaction tube rather than the heating block, resulting in a much closer adherence to the preset conditions of temperature and time. Stringent temperature control is achieved in some thermal cyclers by accurately tailoring the size of holes in the heating block to 0.5 ml microcentrifuge tubes. The disadvantage with this is that not all suppliers make the same size of tube, a fact which has been exploited by the PCR machine manufacturers who exclusively supply tubes which fit—at a price! Blocks which can be controlled by a thermocouple tube may therefore be cheaper to run, as it is not necessary to match tube size with the same degree of accuracy.

There is still quite a wide range in quality in PCR machines so it is important to identify a machine which suits the needs of the laboratory. The 96-well microtiter plate may be an attractive option to a laboratory assessing, for instance, the rate of hepatitis C virus infection in a bank of 10,000 serum samples. However, the risk of cross-contamination is increased by up to four-fold or more compared to the standard rack of microcentrifuge tubes, as each well has up to eight immediate neighbors (Fig. 3). For another application, PCR sequencing from bacterial colonies or bacteriophage plaques, microtiter plates are ideally suited as amplification is arithmetic, reducing the contribution made by small contaminations of nonspecific sequences.

REFERENCES

1. Saiki RK, Gelfand DH, Stoffel S, Scharf S, Higuchi R, Horn GT, Mullis K, Erlich HA. Primer-directed enzymatic amplification of DNA with a thermostable DNA polymerase. Science 1988; 239:487-491.

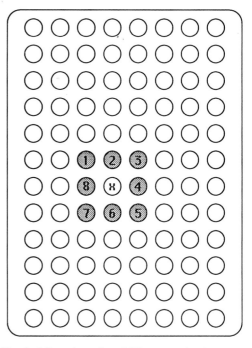

Fig. 3. Microtiter plate PCR contamination.

2. Saiki RK. The design and optimisation of the PCR. In: Erlich HA, ed. PCR Technology. New York: Stockton Press, 1989:7-16.

3. Wang AM, Doyle MV, Mark DF. Quantitation of mRNA by the polymerase chain reaction. Proc Natl Acad Sci (USA) 1989; 86:9717-9721.

4. Frohmann MA. RACE: Rapid amplification of cDNA ends. In: Innis MA, Gelfand DH, Sninsky JJ and White TJ, eds. PCR protocols. San Diego: Academic Press, Inc 1990:28-38.

5. Gilliland G, Perrin S, Blanchard K, Bunn HF. Analysis of cytokine mRNA and DNA: Detection and quantitation by competitive polymerase chain reaction. Proc Natl Acad Sci (USA) 1990; 87:2725-2729.

6. Innis MA, Gelfand DH. Optimization of PCRs. In: Innis MA, Gelfand DH, Sninsky JJ, White TJ, eds. PCR protocols. San Diego: Academic Press, Inc, 1990: 3-12.

7. Kawasaki ES. Amplification of RNA. In: Innis MA, Gelfand DH, Sninsky JJ, White TJ, eds. PCR protocols. San Diego: Academic Press, Inc, 1990: 21-27.

8. Singer-Sam J, Robinson MO, Bellve AR, Simon MI, Riggs AD. Measurement by quantitative PCR of changes in HPRT, PGK1, PGK2, APRT, MTase and Zfy gene transcripts during mouse spermatogenesis. Nucleic Acids Res 1990; 18:1255-1259.

9. Gelfand D. Taq DNA polymerase. In: Erlich HA, ed. PCR Technology. New York: Stockton Press, 1989:17-22.

CHAPTER 3

MANAGING THE METHOD

CALIBRATION

Despite its apparent simplicity, PCR has a relatively large number of associated variables. Managing these potential sources of error may at first seem daunting, but it is well worth spending a few days optimizing conditions for a new set of primers in order to avoid problems later when large numbers of samples may be involved. More often than not, the same range of reagent concentrations may be used for several different primer sets with similar nucleotide compositions.

PCR should be optimized for both efficiency and specificity: an efficient reaction is a rapid one and one which faithfully replicates the DNA sequence of interest. This specificity underpins the success of PCR technology, and since the two factors are interlinked, parameters affecting both need to be optimized. These are listed and discussed below.

MAGNESIUM CONCENTRATION

Taq polymerase enzyme requires free magnesium ions (Mg) as a cofactor, so these must be present in the polymerase chain reaction. Mg is also bound, however, by template DNA, primers and dNTPs, and chelated by compounds such as EDTA which may be used in DNA purification.[1] Higher concentrations of Mg also inhibit Taq polymerase.[2,3] Thus the concentration of Mg must be greater than the total dNTP concentration, which represents by far the greatest binder of the ion in the reaction.

Mg also affects DNA strand denaturation temperature, primer annealing specificity and enzyme fidelity. In particular, the formation of 'primer dimers', where primers anneal nonspecifically to produce a short and easily amplifiable product may be increased at nonoptimal Mg concentrations.[4] Clearly it is not enough simply to add more Mg than dNTP—the Mg concentration must be titrated carefully for each primer set.

For this purpose, it is useful to have in stock a set of buffers already made up (sterilized and aliquoted) to various Mg concentrations. Final reaction concentrations from 1 mM to 5 mM in 0.5 mM increments are a

Polymerase Chain Reaction (PCR): The Technique and Its Applications, written by Rosalind A. Eeles, M.A., M.B., B.S., M.R.C.P., F.R.C.R.; Alasdair C. Stamps, Ph.D.; © 1993 R.G. Landes Company.

useful starting point which usually pinpoints the optimal concentration in a single PCR run (see Fig. 1). If not, smaller increments of concentration can be tried around that of the best result, or even more Mg may be added.

If no Mg concentration optimum can be obtained, the other parameters affecting PCR specificity should be examined.

ANNEALING TEMPERATURE

Many primers will anneal specifically and efficiently if the reaction temperature is briefly lowered to 37°C or 42°C for the annealing step of the PCR cycle (see Chapter 2). However, not all oligo sequences are so accommodating, particularly shorter ones, and bringing the temperature down to this level from the denaturation step at 94°C and then raising it to 72-75°C for polymerization may take some time on the average PCR machine, resulting in a substantial increase in run length over 30 or 40 cycles. The tendency of some primers to anneal non-specifically is reduced at higher annealing temperatures. This is particularly applicable to the formation of primer dimers,[4] which may come to dominate a reaction, using up all the available dNTPs.

The denaturation, or 'melting' temperature (T_m) of DNA in solution is affected primarily by salt concentration of the solution and the nucleotide composition and length of the sequence concerned. For oligonucleotides, length considerations are negligible, and T_m varies according to the following formula:

$$T_m = -16.6\log[Na] + 0.41(\%GC) + 81.5°C,$$
$$\text{where:}$$

T_m = melting temperature (the temperature at which 50% of the DNA is denatured);

[Na] = cationic concentration of the solution;

%GC = percentage of dC and dG nucleotides in the oligo.

Based on Bolton and McCarthy, 1962.[5]

The optimum annealing temperature, T_{ann}, for oligos in solution is 25°C below T_m. Thus for any oligo in a PCR buffer at 50 mM KCl:

$$T_{ann} = -16.6\log(0.05) + 0.41(\%GC) + 81.5 - 25°C,$$
$$\text{or}$$
$$T_{ann} = 0.41(\%GC) + 34.9°C.$$

For a pair of oligos containing 50% dG and dC, T_{ann} will be 55.4°C.

This simple calculation may save a lot of extra calibration runs, as even if it does not pinpoint exactly the right annealing temperature, it provides a good starting point around which to vary temperatures by a few degrees until specificity is obtained.

CYCLE NUMBER

Choosing the correct number of PCR cycles can both save time and reduce problems such as misincorporation of nucleotides. With a relatively abundant DNA target (100 ng of genomic DNA, or 2×10^4 copies of a unique sequence), a 30-cycle PCR should be sufficient to give a strong band on an ethidium bromide-stained agarose gel. For lower quantities of input material, cycle number increases proportionally.

It must be borne in mind that, although PCR is theoretically capable of amplifying a given target by 2^n times, where n is cycle number, in practice amplification efficiency drops in response to a number of factors. At the beginning of a run, low template concentration may reduce efficiency; on the other hand, as PCR products accumulate towards the end of the run, product excess and declining dNTP concentration may have the same effect. Other factors affecting efficiency include Mg concentration, and annealing temperature and time. Mispriming leads to diversion of enzyme activity and substrates to the synthesis of nonspecific products. Quality of input DNA may also be important; residual ionic detergents or chelating agents inhibit polymerase activity.[6]

When calibrating a new reaction, it is worthwhile setting up two or three duplicate reactions (aliquots of the same original reaction) for each Mg concentration or annealing temperature and running these for various numbers of cycles, e.g. 20, 30 and 40. When the quantity of input DNA is uncertain, this will readily ascertain the minimum number of cycles required for the average sample.

STEP DURATION

An outstanding feature of PCR is the extreme rapidity with which the reaction occurs at the elevated temperatures used. The processivity of the enzyme (the rate at which it moves along the template strand incorporating nucleotides into the new strand) is enhanced 100-fold on transition from 20° C to 75° C, up to 150 nucleotides per second.[2,3]

Innis and Gelfand[7] recommend a denaturing step of 96° C for 15 seconds and an

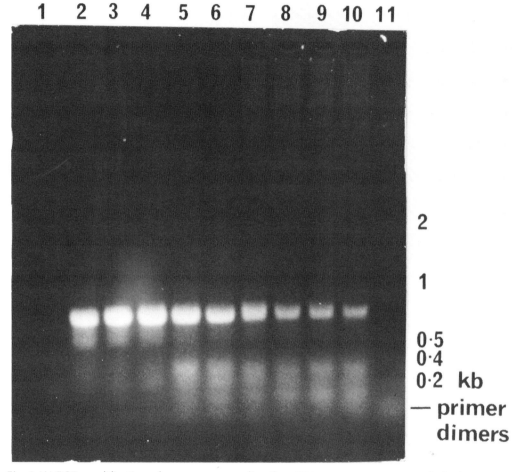

Fig. 1.(A) PCR amplification of sequences encoding the p53 tumor suppressor protein in the presence of varying concentrations of magnesium ions.

Each sample contained: 20 mM Tris-HCl pH 8.4, 50 mM KCl, 0.2 mM dNTPs, 0.05% (v/v) Tween 20, 20 pmol each of two p53-specific 20mer oligonucleotides, 1 U Amplitaq DNA polymerase. Samples 1-10 contained 0.5 mM, 1.0 mM, 1.5 mM, 2.0 mM, 2.5 mM, 3.0 mM, 3.5 mM, 4.0 mM, 4.5 mM and 5.0 mM MgCl$_2$, respectively. Each of these samples also contained cDNA reverse transcribed from 0.1 μg of polyadenylated RNA from normal human fibroblasts. Sample 11 (the negative control) contained 2.5 mM magnesium but no cDNA. Reactions were subjected to 30 PCR cycles using a precalculated primer annealing temperature. The primers were designed to give a specifically amplified fragment of 720 nt. One-tenth of each reaction was applied to an agarose electrophoresis gel. Samples 1-11 were loaded into lanes 1-11, respectively.

p53 cDNA sequences amplified well at magnesium concentrations 1.0 mM or higher in this buffer, but best results were obtained from sample 4 (lane 4), i.e., in 2.0 mM magnesium. Above or below this concentration, specific amplification was sacrificed to the synthesis of nonspecific PCR products. In the negative control, only primer dimers could be detected (lane 11).

annealing time of only 30 seconds, followed by an extension (polymerization) time of 1.5 minutes.

Although the high denaturation temperature would ensure efficient denaturation, it is also highly damaging to the Taq polymerase enzyme, which has a half-life of around 20 minutes at 96° C.[4] Hence short denaturation times are chosen which may not allow for full denaturation, particularly as

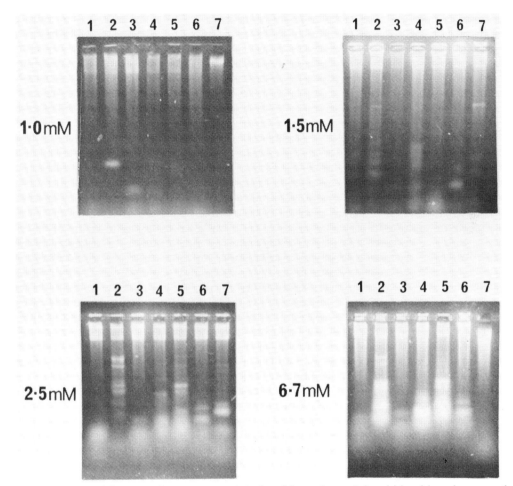

Fig. 1.(B) PCR amplification of genomic DNA isolated from the peripheral blood lymphocytes of a leukemia patient.

Using the buffer and primer concentrations described for Fig. 3(A), and magnesium concentrations as indicated at the left of each panel, DNA was amplified by one oligo specific to a single J_H sequence of the immunoglobulin heavy chain gene region, in combination with each of six different V_H-specific oligos: lane 1-V_H6, lane 2-V_H5, lane 3-V_H4B, lane 4-V_H4A, lane 5-V_H3, lane 6-V_H2, lane 7-V_H-1.

Southern blot hybridization of this DNA had revealed a single immunoglobulin rearrangement involving the J_H sequence, suggesting a single clonal expansion. PCR was used to identify the corresponding V_H region.

Specific amplification was only obtained at 1.0 mM magnesium (top left panel) using the V_H5-specific oligo (lane 2), indicating that this variable region was involved in the leukemic expansion. The other panels demonstrate a dramatic breakdown in amplification specificity with even a small increase in magnesium concentration. In fact, specific amplification was only obtained after extensive calibration of temperature and magnesium concentration.

[Photographs and relevant information reproduced by kind permission of Dr. S. Height (Section of Academic Hematology, Institute of Cancer Research, London, U.K.).]

product concentration rises. Heating most templates to 94° C for 1 minute is sufficient to obtain excellent results and is well within the half-life of the enzyme at this temperature, even for a 40- or 50-cycle PCR.

Annealing times may have an effect on primer specificity, i.e., longer annealing times, especially at lower temperatures, may encourage nonspecific priming. 'Snap-back' of poorly denatured templates to their double-stranded conformation will also tend to be more frequent. At the high primer concentrations used in PCR (many thousand-fold excess in the initial stages of a PCR run), annealing is probably instantaneous for a large proportion of the template present, and 30 seconds provides ample time for efficient annealing in most cases. However, there are situations in which a longer annealing time at the calculated T_{ann} is justified in the first 5 to 10 cycles, such as when degenerate primer sequences are being used to isolate related genes or site directed mutagenesis is being carried out. In this early phase of a PCR run, the original input DNA is still a major source of template, and an extended annealing time ensures better priming. In these cases, a slower ramp time down to the annealing temperature may also enhance priming efficiency.

Extension times, while being extremely rapid, never match the 150 nucleotides per second reported for Taq polymerase activity assays.[3] As described in Chapter 2, 1 minute at 72°C is required for efficient amplification of sequences up to 1000 nucleotides in length. Polymerization times increase with greater DNA fragment sizes, e.g., amplification of a 1500 nucleotide sequence often requires 1.5 minutes at 72° C for the polymerization step.

TEMPLATE LENGTH

The length of DNA sequence which may be amplified by PCR does appear to be limited; templates up to 10,000 nucleotides in length may be amplified. Since most PCR applications are aimed only at the detection of sequences, template lengths of over 1,000 nucleotides are largely irrelevant, but it is worth noting that short fragments are usually much

more efficient PCR templates than long ones. While this means that it is easier to amplify short DNA sequences, it also means that longer sequence amplifications are more prone to nonspecific artifacts caused by mispriming on the original DNA. Once a misprimed sequence has been copied once, it is as easily primed as the correct sequence since it will now have complementary sequences to each primer at its ends. Thus more stringent conditions may be required for longer PCR products.

DETERGENTS

The inclusion of nonionic detergents, e.g., Tween 20, Nonidet-P-40, at low concentrations (see Chapter 2) is known to reverse the inhibitory effects of ionic detergents such as SDS (sodium lauryl sulphate)[8] which are often used in the isolation of genomic DNA from tissue and cultured cells.[1] Nonionic detergents may also help to inhibit proteolytic enzymes which may persist in certain DNA preps, and to which Taq polymerase is particularly sensitive. This is demonstrated in Figure 2, in which recombinant plasmid clones are identified from transfected bacterial colonies by simply adding a small portion of each colony to a PCR tube and subjecting them to 15 cycles in the presence of primers appropriate for the identification of the cloned sequence. In the absence of detergent, no PCR products are obtained, while the addition of 0.1% (v/v) Triton-X-100 yields a clear amplified product. No ionic detergents are involved in this procedure as the bacteria are lysed efficiently by the denaturation step.

Since nonionic detergents have no discernible effect on PCR efficiency, it is useful to include one in all buffers in order to avoid potential problems associated with ionic detergents or proteolytic enzymes.

PRIMER CONCENTRATION

If the concentration of oligos in a reaction is high, nonspecific priming may be promoted. Most researchers using PCR include 100 pmol of each oligo per 100 μl reaction. Reducing this quantity to 20 pmol of each oligo has little effect on yield and may increase specificity in some cases. In

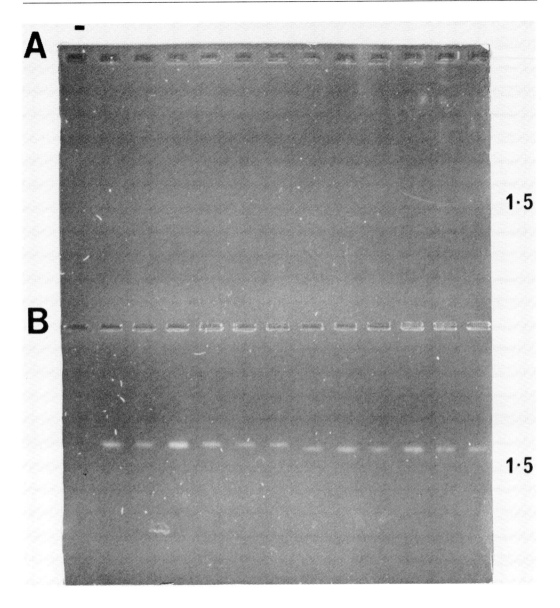

Fig. 2. Direct PCR amplification from individual bacterial colonies.

PCR products amplified from normal human fibroblast cDNA using oligos specific to the coding region of the p53 gene were inserted into a plasmid vector which was then introduced into E. coli. The resulting plasmid-containing bacterial colonies were individually dispersed in 50 µl of liquid medium and incubated at 37°C for 1 hour. After vortexing to resuspend the bacteria 1 µl of each suspension was added to 24 µl of a solution containing 20 mM Tris-HCl pH 9.0, 50 mM KCl, 0.2 mM dNTPs, 2.25 mM MgCl₂, 5 pmol each of two oligos specific to recombinant plasmids, 1 U Taq polymerase and either no detergent (A), or 0.1% (v/v) Triton-X-100 (B). The reactions were then subjected to 15 cycles of PCR at the precalibrated optimum annealing temperature, and 5 µl applied to an agarose electrophoresis gel. A negative control reaction containing no bacteria was included (left hand lane). In the absence of detergent, no amplificatin was observed (panel A) while the buffer containing 0.1% Triton-X-100 produced easily distinguishable PCR products of the correct size (panel B). The slight size differences observed between bands in the first six and last six lanes are due to a 24 nt modification used in one set of the original amplifying primers.

addition, five times as many reactions can be carried out with the same stock of oligos.

CONTROLS

The amplification potential of PCR places stringent requirements upon control reactions to ensure the validity of each experiment.

Two control reactions are necessary for each PCR run. The first, and most important, is the negative or 'no DNA' control. This confirms the DNA-free status of all solutions and apparatus involved in setting up PCR. It contains all the reagents used in the experiment, with the exception of sample DNA, which is replaced by an aliquot of the solvent in which the sample DNA is dissolved. Very often the negative control, containing no other nucleic acid substrate, produces more primer dimers than a positive reaction. The diffuse band seen at the bottom of many PCR gels at a position corresponding approximately to the combined molecular weight of both primers may therefore be somewhat more intense in the negative control than in the sample reactions. This is not due to DNA contamination, which is signified by the appearance in the negative control of a band equivalent in size to the predicted distance between the two primers on the DNA template.

Newly made up and aliquoted samples should be extensively tested in negative control reactions to screen for 'sporadic' positive reactions.[9] These occur when DNA contamination is dilute enough to be on the detection limit of the PCR conditions and reagents used. In this situation, small variations in efficiencies of individual PCRs (due to normal pipette error and slight variations in heating block accuracy) show up as occasional false positives in a series of negative samples. In a reaction system set up to detect single molecules of DNA, sporadic false positives may result from very low level contamination in which the probability of a single fragment of specific DNA template being present in one reaction volume of reagents is less than unity.

Having obtained negative results on all reagents, it is important in certain situations, e.g., diagnostic PCR, to identify a series of good positive controls. A 'good positive control' in PCR is not necessarily one which produces a high intensity band on an ethidium bromide stained gel. On the contrary, such a control only demonstrates the fact that even an inefficient PCR can still be made to work and presents a potential contamination source due to the high level of input DNA (see Chapter 4). PCR positive controls should amplify weakly but consistently.[10] This type of control confirms the sensitivity and efficiency of the reaction.

Choice of positive control depends on the nature of the material to be amplified in the sample reactions. The control should ideally be derived from one of these samples. Dilution of controls derived from clinical material or other sources of genomic DNA should be avoided. It may be that the DNA isolation procedure leaves an inhibitory level of ionic detergent in the samples, which is sufficiently reduced by dilution for its effect to be reversed by the nonionic detergent contained in the PCR buffer. The result of this would be that weakly positive samples failed to amplify, while the diluted positive control consistently gave a good result. A more accurate positive control would be provided by an undiluted sample which had been repeatedly shown to be weakly positive in the presence of negative controls.

Positive controls should, if possible, be compared to a dilution series of a plasmid of known concentration which contains the sequence of interest (Fig. 3). This plasmid should be subjected to a mock 'isolation' procedure to give an accurate assessment of the sensitivity of the reaction. This comparison will demonstrate the detection limit of each PCR system.

DESIGNING OLIGOS

The specificity of PCR results from the use of oligonucleotides designed to anneal precisely to a single DNA sequence. Two short sequences are chosen which are separated by a known sequence length and which are complementary to opposite strands of this sequence. Which sequences are chosen will be influenced by a number of factors.

TEMPLATE SEQUENCE LENGTH

For the straightforward detection of DNA sequences, the primers must be placed far enough apart on the template to obtain a distinct band on a gel, but not so far apart as to create inconveniences such as increased PCR run and gel electrophoresis time. Two-hundred fifty nucleotides is a useful size for these types of PCR product, as it is distinct from the size of a primer dimer but short enough to observe after brief agarose electrophoresis (1 hour) and to sequence through completely on one sequencing gel.

UNIQUENESS

Obviously each primer should have a unique sequence to prevent mispriming on a nonspecific DNA sequence. Usually PCR primers are between 18 and 28 nucleotides long, sufficient to ensure uniqueness of sequence if derived from nonrepetitive DNA, and to avoid mispriming if a high enough annealing temperature is used. It may give some reassurance to compare each primer with all known sequences on a database of the organism concerned, but since most of the genome of even the most well-characterized eukaryotes is

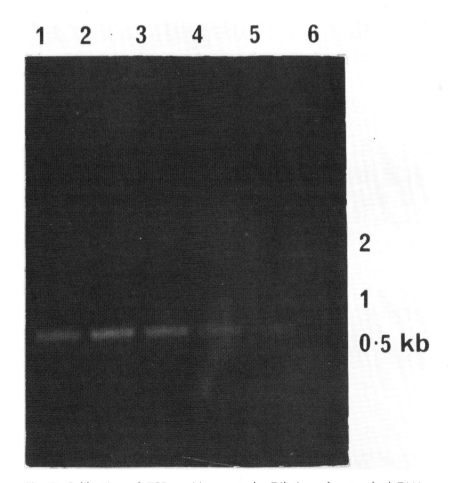

Fig. 3. Calibration of PCR positive controls. Dilution of a standard DNA sequence of known concentration gives a semi-quantitative assay of the detection limit of the method and provides a calibration series for comparison to a typical postive sample. In this example, lanes 1-5 contain PCR products obtained from 10^7, 10^6, 10^5, 10^4 and 10^3 molecules of specific sequence, respectively, by 30 cycles of amplification in the presence of appropriate oligos. Lane 6 is a negative control containing no input DNA.

presently unknown, there seems little point in wasting time on this.

LACK OF INTER-OLIGO COMPLEMENTARITY

Complementary sequences between oligos will encourage the formation of primer dimers, particularly if they occur at the 3' ends of the oligos. These can be readily detected using standard computer sequence homology programs - the complementary sequence of one primer is screened for homology to the sequence of the other primer. If complementarity is found, the position of one primer should be shifted along the proposed template until a more suitable sequence is found.

LACK OF INTERNAL COMPLEMENTARITY

Similarly, complementary regions within the sequence of one oligo may cause problems in PCR due to secondary structure formation - the oligo 'folds up' on itself, preventing annealing or even promoting annealing to the wrong sequence (Fig. 4). Some computer sequence packages provide a program which detects internal complementarity. Homology comparison of the oligo with its complementary sequence will also highlight such regions.

dG + dC CONTENT

A 50% dG + dC content should be aimed for if possible. dG and dC bind more efficiently to complementary DNA, raising the annealing temperature and enhancing specific annealing. A higher frequency of dG and dC at the 3' end of the oligo should therefore stabilize annealing at this end, promoting Taq polymerase activity. However, it is not always possible to obtain this 'ideal' PCR oligo sequence, and in practice oligos with lower dG + dC contents may do just as well.

AVOIDING LONG RUNS OF ONE NUCLEOTIDE

These appear to be a disadvantage in PCR primers, although this may have more to do with problems in synthesis and purification of the oligo than with efficient annealing. If problems are experienced when such a primer is involved, it may be analyzed by running 1 or 2 µg on a polyacrylamide gel. On ethidium bromide staining and UV light exposure, most of the fluorescence should be restricted to one band. If a ladder of bands is observed, the oligo is contaminated by subfractions and must be purified by polyacrylamide gel electrophoresis or HPLC. In general, better results are obtained with HPLC-purified oligos.

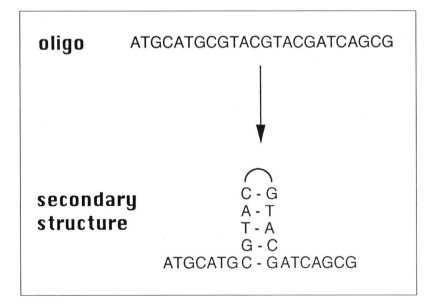

Fig. 4. Intrastrand complementarity.

Computer programs are currently available which aid PCR primer design, and these can be particularly useful when designing degenerate oligos for the isolation and identification of related gene sequences (see Chapter 6: Isolating Genes by PCR Amplification with Degenerate Primers). In the design of oligos for 'standard' PCR, however, they rarely provide any significant advantage over the use of the above rules and the formula for defining annealing temperature.

PROBLEMS AND PITFALLS

PCR PRODUCT CONTAMINATION

The danger of the fact that PCR is such a powerful amplification technique is that contamination can occur. This is the most serious problem with PCR since large numbers of copies of amplified sequences are obtained and even minimal carry-over can result in contamination problems. Such contamination at the very least ruins an experiment, but at worst is a disaster in a diagnostic laboratory. An analogy for the potential scale of the problem is as follows: if the products of a 100 µl PCR were added to an olympic size swimming pool then a 100 µl

aliquot from this pool would contain 40 amplifiable molecules![9] Of course, theoretically, only one of these molecules would be needed for a further PCR. The following can be used to try to reduce contamination:

1.) Physical separation of pre- and post-PCR amplifications. Designated areas for handling PCR products should be separated from those where PCR is set up. There should be a separate set of supplies and pipetting devices dedicated for the specific handling of PCR products and gloves should be changed frequently, particularly when moving from the post- to the pre-PCR product room.

2.) Positive displacement pipettes or plugged pipette tips. The end of the barrel of the pipettes can be contaminated by aerosols during pipetting. The ease with which this occurs can be demonstrated by pipetting a radioactive solution and monitoring the pipette barrel-end with a Geiger counter. This source of contamination can be minimized by using postive displacement pipettes or pipette tips with plugging material in the tips (see Fig. 5). In our experience, the latter plugged pipette tips are easy to use and completely remove contamination. They are in use routinely in our p53 diagnostic laboratory.

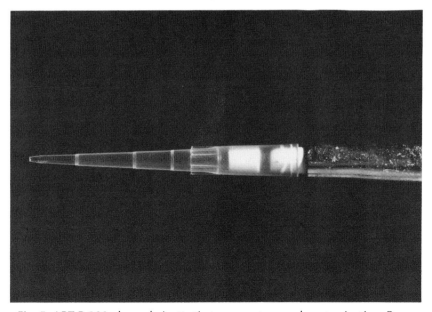

Fig. 5. ART P-200 plugged pipette tip to prevent aerosol contamination. Reproduced with kind permission of NBL, UK and manufactured in USA by Moleuclar Bio-Products Inc., USA.

3.) Aliquoting of reagents. All reagents should be kept frozen in the pre-PCR room and only aliquots should be thawed each time for the setting up of PCR reactions. Aliquoting also ensures that if contamination occurs only a small amount of reagent has to be discarded.

4.) The use of negative controls. These should always be included in each PCR experiment to monitor potential PCR contamination and, of course, the control reactions should be set up last (Fig. 6)

If a contamination occurs

The best way to deal with contamination is to prevent it occurring, but if pipettes do become contaminated they can be cleaned by soaking in 1 molar hydrochloric acid to depurinate any residual DNA. UV or gamma irradiation can be attempted to remove contamination,[11,12] but if aliquots of reagents become contaminated it is far better to discard them than to attempt decontamination.[12] Varshney et al[13] have developed a chemical method to prevent carry-over of PCR product. They include dUTP in the first PCR. This can be degraded by using a deoxyuridine-targetted enzyme reaction. Uracil-N-glycosylase cuts DNA strands at alkaline pH and high temperatures.

OPTIMIZATION OF THE PCR

MAGNESIUM CONCENTRATION

Alteration of the magnesium concentration can effect primer annealing strand association temperatures of the DNA. Product specificity, the formation of primer dimer artifacts (where primers anneal to each other rather than to the template) and enzyme activity and fidelity are also affected by the free magnesium concentration. Taq DNA polymerase needs free magnesium which is not bound to the DNA or dNTPs. In general, PCRs should contain 0.5-3.0 mM magnesium (Fig. 1).

DEOXYNUCLEOTIDE TRIPHOSPHATE CONCENTRATION

The optimal concentration is between 20 and 200 µM and the 4 dNTPs should be used at equivalent concentrations to minimize misincorporation errors.[14] If one is at a higher concentration than the others, it will be preferentially incorporated.

THE PRIMERS

The primer concentrations should usually be between 100 and 500 µM. Higher concentrations can encourage nonspecific annealing and product formation and can result in the formation of primer dimers where primers anneal to themselves and not the template. This is seen

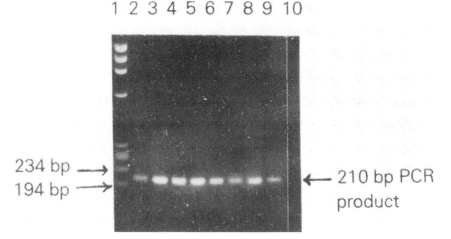

1 2 3 4 5 6 7 8 9 10

234 bp →
194 bp →

← 210 bp PCR product

Fig. 6. Ethidium bromide—stained 2% agarose gel. PCR products, each of 210 base pairs in lanes 2-9. Lane 10 is a negative control (no DNA to demonstrate that there is no PCR contamination). Lane 1: φ x 174 bacterial digested DNA, digested into fragments of known size to act as a size marker. (Reproduced with kind permission of Elsevier. From Eeles et al. Eur J Cancer 1992; 28:1.)

on analysis of the PCR product as failure of product synthesis and a strong band of the combined length of the primers.

DENATURATION TEMPERATURE AND TIME

Typical denaturation conditions are 94-95° C for 30 seconds to 1 minute. However, higher temperatures are needed for GC-rich templates. Incomplete denaturation of the target template and PCR product is a common cause of failure of the PCR. The temperature is ideally monitored inside the reaction tube by using a thermocouple which should be filled with mineral oil to prevent evaporation; the volume should be that in the PCR reaction. Some PCR machines require the thermocouple tube to be in a certain position for optimum temperature control.

ANNEALING TEMPERATURE

The annealing temperature and primer annealing time depend on the length, concentration and base composition of the primers. Usually the annealing temperature will be 50-60° C. (See Chapter 2.) Increasing the annealing temperature reduces the chances of nonspecific priming and thereby nonspecific product formation.

TIPS FOR ORDERING PRIMERS

Typical primers are 18-25 nucleotides long and should have about 50% dG plus dC composition. As a rough rule of thumb the calculated T_mS for a given primer pair should be balanced and should be about 2° C for an A or T and 4° C for a G or C.[15] Complementarity between the primers should be avoided as this encourages primer dimer formation and the primer should not have complementary sequences within itself as this will encourage secondary structure within the primer, reducing its annealing efficiency. Sometimes substitution of 7-deaza-2'-deoxy GTP for dGTP is useful to reduce the formation of secondary structure. The use of 7-deaza dGTP in the PCR also reduces the formation of compressions in subsequent sequencing reactions.[16]

PRIMERS WITH STRETCHES OF PURINES OR PYRIMIDINES SHOULD BE AVOIDED

When ordering primers, these are written 5' to 3' and the upstream primer will be written as per the sense strand of the DNA and the downstream primer complementary to the sense strand, 5' to 3', i.e., towards the upstream end (Fig. 7).

```
ACACAACTGTGTTGACTAGCAACCCAG.......TGTGGGGCAAGGTGAACGTGGATGAAGTTG
                   ************************CCACTTGCACCTACTTCAAC



The Primers are:

ACACAACTGTGTTGACTAGC --------->
5'end                 3'end



                         <----------CCACTTGCACCTACTTCAAC
                                    3'end          5'end


        When ordering the primers, write as:

        5'ACACAACTGTGTTGACTAGC

        5'CAACTTCATCCACGTTCACC
```

Fig. 7. *The β globin gene sequence with sequence to be amplified by PCR in bold. Complementary sequence to be amplified is starred. Primers are written in the same direction as sequence is created by Taq DNA polymerase, i.e., 5' to 3'.*

NESTED PRIMERS

These are primers which are 3^1 to the initial pair of primers. The first PCR is done with the outermost or 5^1 primers and then the PCR is repeated using the PCR product as a template with nested primers. This markedly increases the specificity of the PCR but, of course, care should be taken to avoid contamination.

TEMPLATE CONCENTRATION

If the template concentration is too low or too high, nonspecific amplification or inhibition of the reaction respectively will occur. The ideal genomic DNA concentration is about 100 ng for a 50-100 μl PCR reaction.

THE QUALITY OF THE TEMPLATE

DNA in paraffin-embedded sections tends to be more degraded than genomic DNA, and PCR of products much larger than 400 base pairs can be difficult. Inhibitors also tend to be present in paraffin embedded sections, and it has been our experience that the more dilute the DNA from this source the better the PCR product. Phenol-chloroform extraction also tends to yield poorer results than PCR of DNA from paraffin sections which has been extracted using simpler methods such as extraction with detergent and proteinase K and PCR of the supernatant directly without the phenol-chloroform extraction. Amplification from paraffin-embedded tissue is less efficient than from blood or frozen tissue and to compensate for this, cycling parameters can be altered by increasing the number of cycles, e.g., from 30 to 40, and increasing the time at each step in the cycle (such as from half a minute to a minute). Special digestion buffers are available for DNA extraction from paraffin sections fixed in Bouin's fixative which until recently resulted in almost complete failure of PCR amplification.[17]

THE PLATEAU EFFECT

The initial PCR cycles have an exponential rate of product accumulation. The plateau effect describes the attenuation in this exponential rate of product production which occurs during later cycles. Several conditions can effect the plateau:

1. the utilization of substrates, either primers or dNTPs
2. the stability of the reactants
3. end product inhibition
4. competition for reactants by nonspecific products or primer dimers
5. reannealing of product at higher concentrations which prevents the extension process
6. incomplete denaturation at higher product concentration.

The problem with the plateau effect is that it encourages nonspecific amplification and so increasing number of cycles does not increase specificity or efficiency of the PCR.

THE TAQ POLYMERASE ENZYME

The recommended concentration range for Taq DNA polymerase (Perkin Elmer Cetus) is between 1 and 2.5 units[18] per 100 μl reaction. If the concentration of the enzyme is too high, then nonspecific amplification will occur. The introduction of Taq polymerase enabled longer products to be amplified. Klenow enzyme can only amplify products up to 400 base pairs whereas Taq polymerase can amplify up to 10 kb.[19] Although Taq polymerase is more stable at higher temperatures than Klenow and therefore new enzymes do not have to be added at each cycle, the enzyme should still be stored at -20° C and only taken out of the freezer when one is ready to add it to the PCR reactions.

POLYMERASE ERROR RATE

The error rate of Taq polymerase is about 3 per 1000 nucleotides; it is this high because Taq polymerase lacks proofreading exonuclease activity. New polymerases are now available which contain this activity and therefore have a higher fidelity.[20] Since DMSO and formamide both decrease the activity of Taq polymerase, they have been used to reduce nonspecific amplification in PCR reactions when added in small concentrations, for example, 10% DMSO or 1.25 to 5% formamide.[21-23]

IMPROVING YIELDS OF LONG PCR PRODUCTS

When PCR of longer products is attempted, nonspecific priming and synthesis of shorter products can occur in preference. Schwarz[24] has developed a technique where a single strand binding protein is bound to the primers and prevents nonspecific annealing before the annealing temperature is reached. When the latter occurs, the protein disassociates and allows annealing to the correct template.

A 'hot start' is the term used for the reserving of the addition of the enzyme until after a denaturation step of about 5 minutes. This also improves yield.

The use of a tricine buffer without any KC1 improves yield of fragments 3-6kb in size.[25]

REFERENCES

1. Sambrook J, Fritsch EF, Maniatis T. Molecular cloning - a laboratory manual. USA: Cold Spring Harbor Laboratory Press. 1989.
2. Innis MA, Myambo KB, Gelfand DH, Brow MD. DNA sequencing with Thermus aquaticus DNA polymerase and direct sequencing of polymerase chain reaction-amplified DNA. Proc Natl Acad Sci (USA) 1988; 85:9436-9440.
3. Gelfand D. Taq DNA polymerase. In: Erlich HA, ed. PCR Technology. New York: Stockton Press 1989:17-22.
4. Saiki RK. The design and optimization of the PCR. In: Erlich HA, ed. PCR Technology. New York, New York: Stockton Press 1989: 7-16.
5. Bolton ET, McCarthy BJ. A general method for the isolation of RNA complementary to DNA. Proc Natl Acad Sci (USA) 1962; 48:1390.
6. Gelfand DH, White TJ. Thermostable DNA polymerases. In: Innis MA, Gelfand DH, Sninsky JJ and White TJ, eds. PCR Protocols. San Diego, California: Academic Press, Inc 1990:129-141.
7. Innis MA, Gelfand DH. Optimization of PCRs. In: Innis MA, Gelfand DH, Sninsky JJ, White TJ, eds. PCR Protocols. San Diego, California: Acad Press, Inc. 1990:3-12.
8. Saiki RK. Amplification of genomic DNA. In: Innis MA, Gelfand DH, Sninsky JJ, White TJ, eds. PCR Protocols. San Diego, California: Academic Press, Inc. 1990:13-20.
9. Kwok S. Procedures to minimize PCR product carry-over. In: Innis MA, Gelfand DH, Sninsky JJ and White TJ, eds: PCR Protocols. San Diego, California: Acad Press, Inc. 1990:142-145.
10. Kwok S, Higuchi R. Avoiding false positives with PCR. Nature 1989; 339:237-238.
11. Deragon JM, Sinnett D, Mitchell G, Potier M, Labuda D. Use of gamma irradiation to eliminate DNa contamination for PCR. Nucleic Acids Res f1990; 25:18:(20)6149.
12. Prince AM, Andrus L. PCR: How to kill unwanted DNA. Biotech 1992; 12:(3)358-60.
13. Varshney U, Hutcheon T, van de Sande JH. Sequence analysis, expression and conservation of Escherichia coli uracil DNA glycosylase and its gene (ung). J Biol Chem 1988; 263:7776-77794.
14. Zhang W, Deisseroth AB. Effect of nucleotide concentration on specificity of sequence amplification. Biotech 1991; 11:(1)60-62.
15. Thein SL, Wallace RB. The use of synthetic oligonucleotides as specific hybridization probes in the diagnosis of genetic disorders. In: Human genetic diseases: A practical approach. Davis, KE, ed. Herndon:IRL Press: 1986:33-60.
16. Fernandez-Rachubinski F, Murray WW, Blajchman MA, Rachubinski RA. Incorporation of 7-deaza dGTP during the amplification step in the polymerase chain reaction procedure improves subsequent DNA sequencing. DNA Seq. 1990; 1:(2)137-40.
17. Stein A, Raoult D. A simple method for amplification of DNA from paraffin-embedded tissues. Nucleic Acid Res 1992; 20:(19)5237-8.
18. Lawyer FC, Stoffel S, Saiki RK, Myambo K, Drummoned R, Gelfand DH. Isolation, characterization and expression in Escherichia coli of the DNA polymerase gene from Thermus aquaticus. J Biol Chem 1989; 264:6427-6437.
19. Jeffreys AJ, Wilson V, Neumann R, Keyte J. Amplification of human minisatellites by the PCR: Towards DNA fingerprinting of single cells. Nucleic Acids Res 1988; 16:10953-10971.

20. Lundberg KS, Shoemaker DD, Adams MW, Short JM, Sorge JA, Mathur EJ. High fidelity amplification using a thermostable DNA polymerase isolated from Pyrococcus furiosus. Gene 1991; 108:(1)1-6.

21. Sarkar G, Kapelner S, Sommer SS. Formamide can dramatically improve the specificity of PCR. Nucleic Acids Res, 1990; 18:7465.

22. Comey CT, Jung JM, Budowle B. Use of formamide to improve amplification of HLA DQ alpha sequences. Biotech 1991; 10:(1) 60-1.

23. Zhang W, Hu GY, Deisseroth A. Improvement of PCR sequencing by formamide. Nucleic Acids Res, 1991; 19:(23)6649.

24. Schwarz K, Hansen-Hagge T, Bartram C. Improved yields of long PCR products using gene 32 protein. Nucleic Acids Res 1990; 18:1079.

25. Ponce MR, Micol JL. PCR amplification of long DNA fragments. Nucleic Acids Res 1992; 20:(3)623.

CHAPTER 4

THE PCR ENVIRONMENT

The polymerase chain reaction is capable, at its greatest sensitivity, of amplifying sequences contained in a single molecule of DNA. This capability confers both wide-ranging analytical and diagnostic potential (from forensic analysis[1] to the detection of inherited disease in a single cell of a pre-implantation embryo[2]) and an inherent tendency to produce false results if sufficient precautions are not taken to prevent contamination of samples. A dramatic analogy has been used to illustrate the potential problems contamination may cause:[3] If a typical 0.1 ml PCR reaction generating 10^{12} molecules of amplified DNA were to be uniformly diluted in an Olympic swimming pool of water a 0.1 ml aliquot of this water would still contain 40 amplifiable molecules. The same analogy applies to any concentrated solution of DNA containing only one sequence, such as plasmid preparations, so it is therefore obvious that such solutions must be kept separate from PCR tubes. This is no trivial task: aerosolization of such solutions occurs every time their tubes are opened and every time they are pipetted. Aerosol particles disseminate in air currents around the laboratory, settling on (and contaminating) every surface. Swabs taken from benchtops, refrigerator and door handles, the ends of pipettes, test tube racks, etc. in established laboratories have been shown to contain amplifiable DNA. The problem extends even to the samples themselves, where a tiny drop of liquid transferred from one sample may give a false positive result to another.

This nightmare scenario is avoided if the correct PCR environment is set up before the commencement of PCR-based experiments. Often laboratories set up and use PCR without suitable containment simply because the first few experiments were not contaminated. This is a false sense of security as the level of amplifiable DNA molecules in such an environment gradually rises until there is a very strong likelihood of reaction tube contamination and PCR results start to become unreliable. 'Containment' of PCR is therefore essential, particularly in the diagnostic context. Fortunately, PCR as an exponentially replicating system has long-established precedents in the form of microbiology and tissue culture. The PCR environment is therefore based to a large extent on principles which have been applied for many years in other areas of laboratory research.

Polymerase Chain Reaction (PCR): The Technique and Its Applications, written by Rosalind A. Eeles, M.A., M.B., B.S., M.R.C.P., F.R.C.R.; Alasdair C. Stamps, Ph.D.; © 1993 R.G. Landes Company.

MAINTAINING A NUCLEIC ACID-FREE ENVIRONMENT

THE 'DEDICATED' LABORATORY

The most effective way to prevent contamination of PCR tubes is to set them up in a separate room which is not used for any other experiment involving naked DNA or cellular material (tissue, cultured cells or bacteria). This is the 'nucleic acid-free environment'. Many will argue against such stringent measures, saying that good laboratory practice should be sufficient to prevent contamination. However, a laboratory applying PCR to a particular DNA sequence will accumulate copies of this sequence at an exponential rate in stored sample tubes. Every time a tube containing PCR products, or concentrated solutions of plasmids with inserts of the same sequence, is opened, an aerosol is formed disseminating millions of copies of the sequence into the laboratory. These tiny droplets come to rest on every surface—benchtops, test-tube racks, pipettes, laboratory coats, etc. After a few weeks' plain sailing, negative control reactions inexplicably turn positive, and no PCR run is now completely reliable. In any situation, clinical or research, this is unacceptable. The attraction of a separate room to keep polymerase chain reactions away from their products is immediately apparent.

Unfortunately, space is at a premium in many laboratories and it can be very difficult to obtain and keep control of any separate room, however small, in such a situation. This leads many researchers to settle for a separate corner of their main laboratory, which in the course of time becomes contaminated. In such a situation, a laminar flow hood may be the answer.

THE LAMINAR FLOW HOOD

Although a relatively expensive item of equipment, the laminar flow containment hood is virtually ideal for laboratories in which PCR is a major focus. It forms a PCR workstation which is easy to decontaminate and also a separate laboratory occupying a minimum area, i.e., a room within a room. Airflow within the hood (Fig. 1) ensures that (a) no particles enter the set-up area from the outside and (b) aerosols are carried away

from open tubes lined up in racks placed parallel to the front and back of the hood, i.e., parallel to the laminar flow. Air in the hood flows under the work area through grids in the front and back and is then passed through microfilters and either exhausted or recycled. The microfilters are replaced at regular intervals. After use the work area is wiped over with a sterilizing solution (70% ethanol is sufficient) and the hood is sealed at the front. Many hoods contain a germicidal ultraviolet lamp—leaving this switched on in the sealed hood renders any remaining DNA incapable of subsequent amplification.[3] In the absence of an integral UV lamp, decontamination may be effected by a portable, shortwave (254 nm) UV lamp.

NOTE: *UV lamps should be shielded by perspex or glass and warning notices placed wherever they are operating.*

Obviously, for a laminar flow hood to be most effective, external sources of contamination should be kept away from its immediate vicinity. Such sources include: centrifuges, vacuum driers, plasmid isolations, bacterial cultures, etc. It is convenient to keep the thermal cycler next to the hood as an effective separator.

GOOD LABORATORY TECHNIQUE

Whichever type of area is chosen for setting up PCR, a few straightforward rules should be observed to maintain it as a nucleic acid-free environment.

Common sense underlies most of the principles applied to handling PCR tubes at the point of set-up. Seven basic rules apply to setting up PCR.

1. Every component should be sterile (see Chapter 2).

2. Gloves should always be worn by the operator to reduce the risk of skin-flake contamination. These should also be changed frequently as they may become externally contaminated. Laboratory coats with elasticated cuffs and even plastic mob-caps have been donned in an effort to prevent operator contamination of PCR tubes![4]

3. All tubes containing solutions should be centrifuged for 2-3 seconds before opening to reduce the risk of aerosol formation.

4. Tubes should not be left open longer than necessary. Opening other tubes, or holding or moving them over open tubes is to be avoided as contaminants may fall directly into the reactions.

5. Discarded pipette tips, etc., should go into a container which can be sealed before removal, such as an empty chemical bottle.

6. Nonsample components should be added first. Sample DNA is added last (except in "Hot-start PCR"), and each tube capped before moving on to the next. Negative controls should be set up last.

7. Once the reactions are cycling and everything has been put away, the surrounding surfaces should be wiped over with a sterilizing solution to discourage growth of microorganisms and keep down dust, which

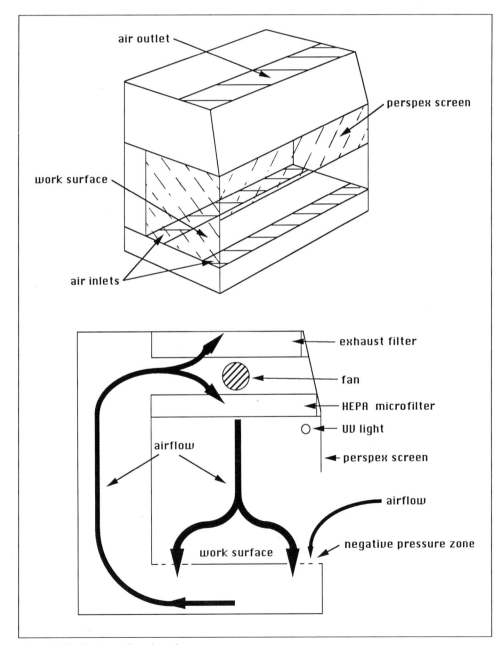

Fig. 4. The laminar flow hood.

could accumulate aerosol-borne DNA. The area should remain clutter-free, as any extra surface is a potential source of contaminating dust particles.

Exposure to short-wave UV light of 254 nanometer wavelength degrades DNA beyond the point at which it is still PCR amplifiable.[4] UV sources are now available which have been specifically designed for PCR laboratories. Many laminar flow hoods are already fitted with sterilizing UV lights which are equally effective in breaking down nucleic acids. 12 hours exposure to one of these UV sources is required to completely destroy all DNA contaminants.

NOTE: *UV lamps may damage eyesight and must be shielded by perspex or glass wherever they are operating. The glass frontage of a laminar flow hood will filter out short wavelength UV light.*

Dedicated apparatus and reagents

Not only should the area where PCR tubes are set up be kept separate from exogenous DNA sources, but setting up must itself be carried out using dedicated apparatus and reagents. That is to say, all automatic pipettes, tubes, tips, racks and solutions must be used only for PCR and never brought into contact with any other possible source of amplifiable DNA. The best way to isolate these items from other potential uses (and users!) is to keep all apparatus in a designated PCR drawer or cupboard and all solutions and enzymes in a separate -20°C freezer compartment. DNA samples must also be kept separate, for example in a sealable box in a refrigerator or cold room. Plastic bottles, such as empty reagent bottles with screw-on lids, are useful for discarding used tips and empty tubes as they ensure containment and facilitate disposal of waste.

Autoclave and aliquot solutions and reagents where possible. Those which cannot be autoclaved (dNTPs, Taq polymerase, oligos) may usually be obtained in purified form, free of nonspecific DNA. All other solutions in the standard PCR mix (see Chapter 2, Table 2) may be autoclaved at standard pressures for 30 minutes to degrade any contaminating DNA.[3] Sterile solutions should be aliquoted in volumes sufficient for the average PCR run

carried out in the laboratory, and batch-labelled. Any contamination occurring during subsequent use is thus confined to one aliquot, while unsuitable batches may also be readily identified and disposed of. EVERY solution used for PCR should be sterile and aliquoted—including water for dilution. The water should be HPLC purified.

PREVENTING CROSS-CONTAMINATION

Gloves

Use disposable gloves when handling any apparatus or reagents which will be involved in setting up reactions. Gloves should ideally be changed regularly while setting up PCR. The wearing of gloves is primarily aimed at preventing contamination of PCR tubes with skin flakes and other particles on the hands of the operator which may carry amplifiable DNA. The gloves may, however, become externally contaminated by aerosols of DNA-containing solutions and necessary or accidental touching of contaminated surfaces. Regular glove changing reduces the likelihood of this contamination being passed on to the reaction tubes.

Tubes

All test tubes used for PCR should previously be sterilized, either by autoclaving or by exposure to gamma radiation (this is carried out by certain manufacturers). Briefly centrifuge tubes containing DNA or oligo solutions. Avoid the use of tubes with caps which are liable to snap open as these create aerosols which may land in reaction tubes. 'Screw-cap' microfuge tubes are ideal for storing PCR solutions. Avoid opening or holding any tubes above open reaction vessels as any contaminating droplets would fall directly into the reaction.

Pipetting

Pipetting is itself a source of aerosols which both emanate from the end of the pipette tip and arise within the space between the drawn up liquid and the barrel of the automatic pipette. Of the two forms of contamination, the resulting presence of amplifiable DNA on the end of automatic pipette is the most persistent and difficult to control, necessitating frequent washing of the pipette barrel in 0.1N HCl. This is tedious and time-consuming and does not prevent internal contamination within a single PCR run. Two better options are available:

Positive displacement pipettes.

These are most suited to highly repetitive experiments in which equal volumes of, for instance, premixed reagents are dispensed into large numbers of reaction tubes. The positive displacement pipette works by drawing up a relatively large amount of liquid into a disposable cartridge and dispensing a preset volume each time the built-in piston is depressed. The disposable cartridge ensures that no contamination of the pipette barrel occurs. There are, however, two disadvantages associated with this apparatus: the first is cost, as the cartridges are very expensive compared to ordinary pipette tips, and the second is the inflexibility imposed by fixed volume dispensing. Of course, dispensing of DNA samples is impractical with a positive displacement pipette as it would result in considerable wastage of both cartridges and, more importantly, sample DNA.

Plugged pipette tips

These are ordinary pipette tips which contain a fiber plug positioned just above the maximum fluid level of the tip (Chapter 3, Fig. 5). This creates an aerosol-impermeable barrier between the liquid and the end of the pipette barrel. Plugged pipette tips are now the transfer vessel of choice for many PCR researchers as they totally prevent contamination of automatic pipettes. Although they are somewhat more expensive than standard tips, this is more than offset by the money saved by cutting down the number of repeat experiments necessitated by contamination. They are also supplied gamma-sterilized, further increasing their convenience.

KEEPING REACTIONS AND PRODUCTS SEPARATE

This is probably the most important rule for the prevention of PCR contamination. As already illustrated in this chapter, the high

product output of PCR also confers high contamination potential. A single aerosol particle of a solution of PCR products may transfer 10^8 to 10^9 copies of amplified DNA. This type of contamination is known as 'carry-over' and is the primary reason for establishing a separate PCR environment. A few simple principles are applied to prevent carry-over:

1. Once reactions are complete, they should not be reopened in the PCR set-up area (unless, of course, they are to be reamplified, in which case their product yield and therefore contamination potential is low anyway).

2. Any apparatus used in the processing of PCR products, e.g., automatic pipettes, centrifuges, vacuum driers, gel tanks and all disposable plasticware, should remain in the main laboratory area (or be kept away from the laminar flow hood).

3. Avoid strong positive controls (also see below). Higher concentrations of input DNA increase the risk of carry-over.

4. If PCR products are to be reamplified, for instance after gel elution, extreme care must be taken to avoid cross-contamination of samples. Alternate wells of agarose gels should be left empty. Positive controls should not be run on preparative gels. Cut each individual gel slice out with a clean scalpel blade. Soak gel apparatus, etc., in a solution of 0.1M HCl between preparative gels to depurinate any residual DNA (remembering to wash this out thoroughly before the apparatus is used).

CONTROLS

The 'no DNA' control in PCR is the monitor of contamination by amplifiable nucleic acids, both exogenous and through cross-contamination, and must be included in every PCR run to ensure the validity of the experiment or test.

Positive controls should be selected judiciously. It is tempting to use samples which give a high yield of PCR product, but high yields are often the result of high input rather than efficient or sensitive PCR amplification. The positive control should give a convincing but moderate amount of product. This controls for sensitivity while reducing the risk of carry-over during set-up. Selection for positive controls is best carried out, if possible, by comparison with a dilution series of a known quantity of specific DNA sequence (see Chapter 3).

TROUBLESHOOTING: PCR CONTAMINATION

If the recommendations described above for the PCR environment are implemented, PCR contamination problems should almost never occur. When they do, the entire PCR system has to be systematically examined to pinpoint the cause.

The most common source of problems is operator error. Setting up large numbers of reactions, taking care over each to avoid aerosols, is mind-numbingly repetitive. The tedium can be reduced by the use of premixed reagents (buffer + dNTPs + primers + enzyme, Chapter 2) and by diluting all DNA samples to the same concentration so that the same volume is added each time. This reduces the number of additions to the reaction vessel to three (premix, DNA, mineral oil).

If reagents are aliquoted in batches, it is a relatively simple matter to set up a series of negative controls in each of which one component of the reaction is replaced by another batch. A contaminated solution may then be identified within a single control experiment for most types of PCR amplification.

If false positive results persist, the contamination must originate from another source. This is typified by sporadic contamination of negative controls and is the usual course of events when large scale PCR is carried out and processed in the same laboratory.

If this is the case, the need for a separate, PCR-dedicated room or laminar-flow hood will be self-evident. If a separate PCR work area becomes contaminated, a thorough cleaning of all surfaces followed by overnight UV light exposure should reduce the background of amplifiable DNA, As already mentioned, even a refrigerator door handle can become a source of PCR-amplifiable DNA. In addition, all disposable plasticware in current use should be replaced and all apparatus (automatic pipette barrels, test tube racks, discard bottles) washed overnight in 0.1N HCl, then rinsed thoroughly in autoclaved deionized water.

False positives may also be the result of aerosol cross-contamination. The use of an aerosol-resistant delivery system (positive displacement cartridges or plugged pipette tips) in a laminar flow hood minimizes the incidence of this problem.

REFERENCES

1. Higuchi R, von Beroldingen CH, Sensabaugh GF, Erlich HA. DNA typing from single hairs. Nature 1988; 332:543-546.
2. Coutelle C, Williams C, Handyside A, Hardy K, Winston R, Williamson R. Genetic analysis of DNA from single human oocytes: A model for preimplantation diagnosis of cystic fibrosis. Brit Med J 1989; 299:22-24.
3. Kwok S, Higuchi R. Avoiding false positives with PCR. Nature 1989; 339:237-238.
4. Kitchin PA, Szotyori Z, Fromholc C, Almond N. Avoidance of false positives. Nature 1990; 344:201.

CHAPTER 5

MEDICAL AND GENETIC APPLICATIONS

INTRODUCTION

The development of the polymerase chain reaction has revolutionized medical and genetic research and diagnosis. In particular, the ability to work with small amounts of material has enabled analyses to be performed on specimens where there is very little tissue present such as those from antenatal tests or DNA from paraffin embedded sections. The speed of PCR has made it a very attractive method for diagnostic laboratories. This chapter describes the medical and genetic applications that have been made possible by the introduction of PCR.

INFECTIONS

The polymerase chain reaction is ideal for detecting the presence of bacterial, fungal or viral pathogens. The principle of detection is to attempt PCR with primers complementary to part of the genome of the pathogen which is not complementary to that of the host. Since, theoretically, only one DNA copy needs to be present for PCR to occur, this technique can be used to detect very small numbers of pathogenic organisms or virus particles. It also negates the need for the culturing of the pathogen which often takes days to weeks; also some pathogens are not capable of in vitro growth and so cannot be detected using a culture approach. PCR can also be performed on very small samples such as fine needle aspirates or a tru-cut biopsy.

VIRAL STUDIES

Retroviruses

The study of retroviruses presents several problems. There may be only a small number of viral copies present within highly complex DNA in the

Polymerase Chain Reaction (PCR): The Technique and Its Applications, written by
Rosalind A. Eeles, M.A., M.B., B.S., M.R.C.P., F.R.C.R.; Alasdair C. Stamps, Ph.D.;
© 1993 R.G. Landes Company.

host cell. As few as one actively infected cell may be present per 10,000 uninfected cells in lymph nodes of infected individuals.[1] Since the replication of the retrovirus includes conversion of RNA to DNA which is then incorporated into the host genome, viral sequences can have homology with some human genes, in particular proto-oncogenes, and this can lead to false results. Viral heterogeneity can present a problem. It arises particularly in RNA viruses and viruses replicating through an RNA intermediate (retroviruses and hepadnaviruses). These contain multiple base mutations, insertions, deletions and duplications.[2] This occurs because the polymerase in viral replication lacks proof reading ability. Another problem is that there are multiple members of viral families (such as human immune deficiency virus [HIV] types 1 and 2).

Detection of Human retroviruses

Retroviruses synthesize a complementary double-stranded circular DNA provirus that integrates into the host genome and remains there for the life of the host cell. The human retroviruses characterized to date are human immune deficiency virus (HIV types 1 and 2),[3-5] and human T cell lymphoma, leukemia viruses types I and II (HTLV types I and II). Although antibody serological assays are often used to detect the presence of the virus, it can be useful to detect the viral pathogen directly, for example, in cases of the acquired immune deficiency syndrome where the patient is still antibody negative[6] or in babies born to HIV positive mothers who will have acquired antibodies across the placenta and who may not be necessarily infected with the virus.[7]

The problems associated with detecting retroviruses include the fact that they have no transcriptional activity in the proviral genome state, there will be a small number of infected cells in the peripheral blood and there will be only a small number of proviral copies per cell infected. There may be multiple members which are related but distinct within the genome and some of the infected cells may be in sanctuary sites such as the brain which are not easily monitored.

Human T cell lymphoma, leukemia viruses

The detection of human T cell lymphoma, leukemia virus type I (HTLV1) proviral sequences in the genome of patients with tropical spastic paraparesis was first achieved using PCR.[8-10] Kwok et al[11] used PCR to detect both HTLVI and II in patients with lymphoproliferative and neurological diseases. It can also be used to detect asymptomatic carriers, but before identifying somebody as a carrier the chance of this result being a false positive must be excluded. Contamination can be reduced as outlined in Chapter 3. If the patient is seronegative, then it is preferable to confirm the results using at least two sets of primer pairs.[12] Some individuals who are infected in utero do not seroconvert until they are in their thirties.[8]

Primers can be chosen that will detect both HTLVI and II;[10,11] alternatively primers can be chosen which will discriminate between the two species.[13] (For primer sequences please see Ehrlich.[14]) PCR was important in showing that the HTLVI virus associated with adult T cell leukemia was the same virus that caused tropical spastic paraparesis[9,15] and has been used to document infection in the individuals presenting with symptoms that do not conform to classic adult T cell leukemia.[16]

HTLVI is a member of the Oncovirinae sub family of retroviruses and is associated with an aggressive type of adult T cell leukemia particularly found in the Caribbean and Japan. HTLVII has about 60% homology in its nucleic acid sequence and was isolated from an individual with a rare T cell variant of hairy cell leukemia.[17,18]

HTLVI is associated with a chronic progressive myelopathy in which there is a degenerative lower limb paralysis (tropical spastic paraparesis), particularly seen in the Caribbean.

HTLVII has been shown to be present in the intravenous drug abusers[19] but has not been associated with any disease therein to date.

HIV types 1 and 2

These are members of the Lentiviridae subfamily of retroviruses and are generally thought to be responsible for the development of acquired immunodeficiency disease syndrome (AIDS) where profound immunosuppression is associated with opportunistic infection and the development of certain tumors such as aggressive non-Hodgkins lymphoma and Kaposi's sarcoma. The two viruses have a 60% homology in their nucleic acid sequence for the *pol* and *gag* genes and 40% for the LTR region and other viral genes. In vitro propagation of the virus can be used to detect its presence but the procedure is labor-intensive and requires stringent laboratory protection procedures. The presence of antibodies does not always equate with the presence of viral infection, as outlined earlier. PCR overcomes these problems and has an advantage because there is a low level of circulating free virus. HIV I has a very heterogeneous genome.[20-22] PCR is therefore based on amplification of the highly conserved regions. Examples of the primer pairs used are SK38 and SK39[23,24] which amplify 115 base pair regions of the gag gene. Primer sequences can be found from Kelogg and Kwok.[25]

PCR has been used to identify individuals who are infected prior to seroconversion[6] to resolve the infection status of individuals with an ambiguous Western blot assay or serology status, to screen neonates[26-29] and to determine the type of virus present.

Again, because of the enormous power of PCR, cross-contamination should be avoided. There maybe as few as 1 viral particle per 10,000 lymphocytes in peripheral blood. Biological hoods equipped with UV lights that are turned on between PCR experiments can clear the PCR preparation area of PCR product (see Chapter 4).

Hepatitis B virus

The presence of hepatitis B virus is normally diagnosed by detection of hepatitis B surface antigen or antibodies to the surface or core antigen using enzyme linked immunoabsorbent assays or radioimmunoassays. In chronic active hepatitis, active viral replication is present, and PCR can detect hepatitis B virus DNA, usually in the liver, but also sometimes in mononuclear blood cells. It is particularly useful where the level of hepatitis B viral DNA is low. Mack and Sninsky 1988[30] have identified a region in the *pol* gene that is conserved and can be used for PCR yielding a fragment of about 110 base pairs. A single step protocol for detection of hepatitus B virus in the serum has been reported by Manzin et al 1991.[31]

Cytomegalovirus

Cytomegalovirus (CMV) can cause a congenital viral infection resulting in mental retardation and sensorineural deafness. In immunocompromised patients, CMV can cause severe life threatening pneumonitis and a retinitis which can occur in AIDS patients and threatens their sight. CMV is a herpes virus and its genomic diversity is unknown; however, the use of a single primer pair[32] will amplify 90% of all wild type isolates. The problem is that a lot of asymptomatic adults are infected with CMV because many infections are latent. However, PCR has been used to show that CMV can be detected in urine specimens in infected children[33] and in AIDS patients[34,35] with a greater sensitivity than in viral culture.

Human papilloma virus (HPV)

Genital human papilloma viruses are a group of many distinct virus types associated with different diseases and cancers.[36] The epidemiology of cervical cancer suggests that it is caused by an infectious agent and many studies, some of which have used PCR, have shown that in particular, HPV type 16 and 18 are associated with cervical dysplasia and carcinoma.[37,38] The genomes of each of the genital HPV types are unique and yet they show areas of homology, particularly within the open reading frames E1 and L1. A set of consensus PCR primers can therefore be designed that will amplify distinct regions of homology. Hybridization analysis can then be performed using viral type-specific probes or an enzyme digestion can be used to yield to unique DNA digestion pattern specific for each virus.

Enteroviruses

Human enteroviruses are from the Picornaviridae family and consist of more than 60 serotypes. They include polio, coxsackie, echo and hepatitis A viruses. They very commonly cause viral infections in children and in developing countries, polio still causes paralysis in 0.4% of children. Coxsackie can cause life-threatening myocarditis in young adults and coxsackie RNA has been detected in myocardial tissues by PCR.[39] The traditional approach to enterovirus isolation is the use of tissue culture which is time-consuming and inefficient. There is great antigenic diversity between the serotypes and often the viral titer is low, particularly in cerebro spinal fluid.[40] Rotbart has therefore developed a PCR protocol using RNA isolation and reverse transcriptase-PCR.[41]

Other viruses

Lassa fever and adenovirus infection have been detected by PCR.[42,43] Hepatitis C is difficult to diagnose serologically and PCR of viral sequences has confirmed infection in some cases.[44]

DETECTING BACTERIAL PATHOGENS

Bacterial pathogens are usually detected by microscopy and culture. However, where the pathogenic load may be low and the sensitivity required is high, such as in environmental monitoring, PCR can be used to amplify the target gene sequence. The use of gene probes can retain the specificity needed to identify the specific pathogens present. This is particularly applicable to the monitoring of water supplies, and since a high level of sensitivity is required, controls should be included in the experiment. These include the presence of a no-DNA control and control nonbacterial DNA such as that from salmon sperm. These ensure high sensitivity and specificity.

SOMATIC GENE MUTATIONS IN TUMORS

As the individual genes responsible for particular genetic syndromes and genes implicated in the development of human cancer are isolated, it has become increasingly important to be able to analyze them for point mutations and other small genetic changes (deletions, insertions, rearrangements) which are undetectable by Southern analysis. To this end a variety of techniques

have been developed for the detection of "point mutations" (reviewed by Cotton[45] and Rossiter et al[46]). However, few comparisons have been made regarding the relative effectiveness of the different methods. One exception is Theophilus et al[47] who compared the use of RNase A, of chemical cleavage and of GC-clamped denaturing gradient gel electrophoresis (DGGE), concluding that GC-clamped DGGE was the most reliable and informative method.

Currently the three most frequently used methods are single-stranded conformational polymorphism (SSCP),[48-50] denaturing gradient gel electrophoresis DGGE[51-52] and modifications of it, constant denaturant gel electrophoresis (CDGE);[53-55] and hydroxylamine/osmium tetroxide chemical cleavage (HOT).[56-58]

We have used SSCP and have performed a blinded study of various p53 mutant samples in collaboration with two other groups (Prosser, Edinburgh, UK (using HOT) and Borresen, Oslo, Norway, (using CDGE). These results are shown in Table 1.

THE SSCP TECHNIQUE

The SSCP technique relies on the secondary structure of single-stranded DNA being altered by a single base mutation. After PCR, the double-stranded product is denatured by heating and electrophoresed in a gel without denaturant so the single DNA strands run according to secondary conformation, not size. Usually, with a wild-type molecule, two bands are seen on the gel, corresponding to each of the single DNA strands of the PCR product. A heterozygous mutation should be characterized by the presence of two bands identical to those in the normal sample and two bands characteristic of each of the strands of the mutant molecule. Additional bands may be observed because some strands can adopt more than one conformation. In this instance, the normal lane would then also contain additional bands, corresponding to these different conformations. A nondenatured normal product should be included to identify the undenatured/renatured product position since reannealing of a proportion of the sample often occurs (Fig. 1). Orita et al[59] claim that the technique can be applied to fragments up to 400 bp in length and they quote a sensitivity of 83% in 12 samples. Sensitivity of the technique can be adjusted by changing the running conditions and composition of the acrylamide gel, and ideally conditions should be determined on known mutations prior to screening. The results in Table 1 show that for p53 test fragments, the optimal conditions are 0.5 x TBE with no glycerol in the gel at 4°C. These conditions had previously been shown to increase the sensitivity of this technique when assessing known mutants in our laboratory and have also been reported by Spinardi et al.[50] Three test conditions are shown in this study to confirm that the previously defined optimal conditions also had the highest sensitivity in a blinded study. Some workers claim that increasing the polyacrylamide concentration in the gel increases the sensitivity[60] but we have not found this to be the case.

The advantages of SSCP are speed and simplicity and that many samples can easily be processed together. In addition, it is able to distinguish homo- from heterozygous mutations. Modifications of the technique are now available to avoid using radioactivity.[61] The main disadvantage is that it does not detect 100% of mutations. A further disadvantage may be that screening is ideally carried out on small fragments (less than 400 bp). In the event of finding a mutation the whole fragment must be sequenced for confirmation. Overall, a 90% detection rate is not a great disincentive for using a fast and simple technique for mutation screening.

THE CDGE TECHNIQUE

The CDGE technique is a modification of the DGGE system first described by Fischer et al, 1983.[51] DGGE relies on strand dissociation of DNA fragments in discrete sequence-dependent melting domains. The

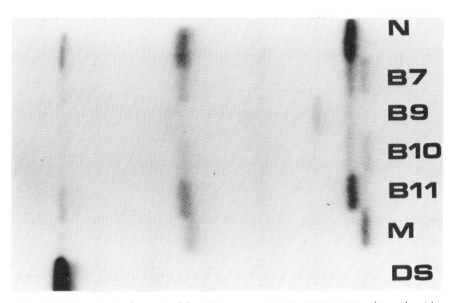

Fig. 1. SSCP analysis of exon 7 of the p53 gene run in 0-5 x TBE, 4.5% polyacrylamide gel at 4°C without glycerol.
DS double-stranded, non-denatured PCR product.
N normal control showing 2 single strands and partially reannealed product. M mutant control
B7,B9,B10 Heterozygous mutant samples with point mutations in exon 7 of p53.
Note the presence of normal and mutant single strands.
B11 normal (Reproduced with kind permission of John Wiley & Sons Inc.)

dissociation causes an abrupt decrease in the mobility of the fragment in polyacrylamide gels containing a gradient of denaturant (urea and formamide). PCR products are more easily analyzed by adding a 40-mer GC clamp (an oligonucleotide of 20 GC runs) to one of the primers, thereby creating a domain with a very high melting temperature at one end of the fragment. All mutations residing in the lower melting domains should theoretically be separated, at least at the heteroduplex level. Since tumor samples are often contaminated with normal DNA, both the wild-type and the mutant, as well as the heteroduplexes of the two, are seen. By estimating the relative intensities of the bands it is possible to determine homo-or heterozygosity of the alleles. The nature of certain mutations can be predicted from the gels since a GC->AT mutation results in the destabilizing loss of a hydrogen bond and slower migration of the sample. An AT->GC change results in a faster migration of the sample.

There are computer programs available from the authors of this technique for analyzing the melting profile of any DNA sequence in order to make educated choices for placing oligos[52] (Fig. 2), but for universal appeal, these programs need to be more user-friendly. The resultant PCR fragment should have only one or two melting domains and the partially melted fragment should achieve a Y shape rather than a bubble. The correct conditions can be chosen either theoretically with the use of computer programs or experimentally with the use of perpendicular gradient gels (Fig. 3) although it is possible that mutations found in the transition regions between two melting domains may present a problem. The fragment sizes screened will depend on the melting profile of the DNA involved and the number of melting domains encompassed, but are usually in the order of 500 bp. In some fragments, suitably placed restriction sites may be useful to remove the lower melting

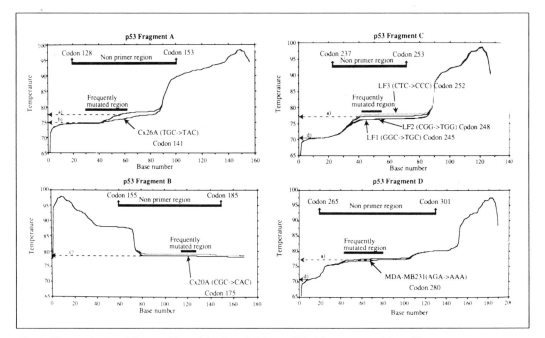

Fig. 2. Theoretical melting profiles of the four PCR-amplified fragments of the p53 gene at temperatures (°C) giving a 50% chance of the base pair being in either double or single-stranded state. The broken lines show the profiles for the different control mutants with localization of the base mutations indicated by arrows. The theoretical melting temperatures a-d correspond to the following experimental denaturant conditions at 56°C in Borresen's CDGE system: a (77-5°C), 50.7%; b (75°C) 42.0%, c (78°C), 51.7%; d (71°C), 32.0%. Figs. 2-5 Reproduced with kind permission of Prof A-L Borresen (Oslo, Norway).

domain(s) sequentially, permitting the analysis of one domain at a time. Disadvantages may include the practical or computer effort required to set up, and, less importantly, the fact that mutations are found in whole fragments with no indication as to their precise location (although perpendicular gels will indicate in which melting domain the mutation resides). Overall, DGGE/CDGE has the potential to pick up 90-100% of mutations without the use of hazardous chemicals or radioactivity, and the simplicity of the eventual routine, especially for screening a number of samples repeatedly across the same gene, makes it worthwhile. CDGE avoids the routine use of a denaturing gradient by experimentally selecting a denaturant concentration at which maximal separation between the wild-type and mutant fragments can be achieved. In addition, the fragments migrate with a consistently different rate through the whole gel, allowing the separation of several centimeters between mutant and wild-type fragments (Figs. 4 and 5).

THE HOT TECHNIQUE

The HOT technique relies on forming a labelled heteroduplex molecule between a wild-type sequence and a potential mutant sequence. The labelling can be of either one or the other or both DNA fragments. The mismatch in the heteroduplex molecule is then chemically modified by either hydroxylamine (HA; modifies mismatched C) or osmium tetroxide (OsO_4; modifies mismatched T) or both, prior to piperidine cleavage at the modified sites. Fragment sizes are then calculated from the position of labelled fragments on a denaturing acrylamide gel (Fig. 6). Having understood the basis of the HOT technique, it is possible to see that it can be carried out at several levels:

1) With both fragments of DNA in the heteroduplex labelled and modification by both HA and OsO_4: This identifies 100% of mutations (and should do so twice over). Effectively, these conditions pertain when testing for heterozygous mutations in samples containing both mutant and wild-type DNA, when the test DNA is labelled and heteroduplexed to itself. In greater than 70 mutations detected in Prosser's laboratory using

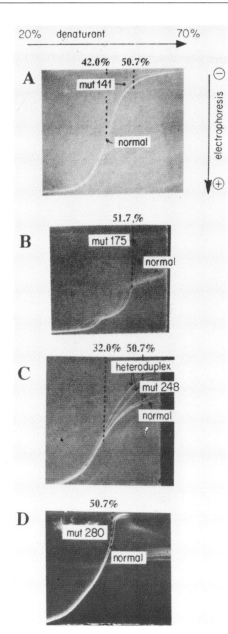

Fig. 3. Perpendicular gradient gel electrophoresis of the four PCR fragments analyzed. A-D correspond to amplified fragments A-D in Fig. 2. DNA from a normal individual was loaded together with DNA from mutant p53 samples Cx26A (TGC→TAC mutation in codon 141) (A), Cx20A (CGC→CAC in codon 175) (B), and MDA-MB231 (AGA→AAA in codon 280) (D), and with DNA from the blood sample LF2 (CCG→TGG in codon 248) (C). The gels were run in a gradient from 20-70% denaturant (where 100% is 7M urea plus 40% formamide). Note the separation of mutants compared with the normal p53 DNA fragment.

the HOT technique, only twice did they fail to see a mutant band (after OsO₄ modification in two samples with a GT mismatch). This technique can sometimes miss GT mismatches, but it should be noted that only a small proportion of GT mismatches are missed and that in any event, the mutations can be detected through HA modification of the reciprocal CA mismatch.

2) With one fragment of DNA labelled (usually, for practical considerations, wild-type DNA) and modification by both HA

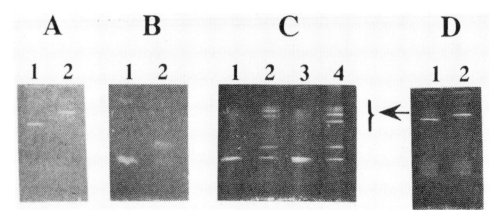

Fig.4. Constant denaturant gels (stained with ethidium bromide) of normal DNA and DNA from six previously identified mutants. Differences in migration reflect differences in the conformation of the DNA fragments. A-D correspond to amplified fragments A-D in Fig. 1. The gels were run at 50.7% denaturant for A,C and D, and at 51.7% for B. Lanes 1, normal DNA (A-D); lanes 2, DNA from Cx26A (A), DNA from Cx20A mixed with normal DNA (B), DNA from blood of LF2 (C), and DNA from MDA-MB231 (D); lane 3, DNA from blood of LF3; lane 4, DNA from blood of LF1. Heteroduplexes are indicated by an arrow.

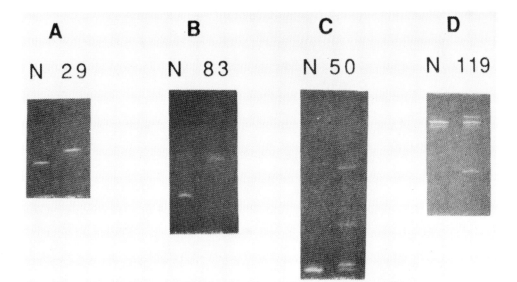

Fig. 5. Constant denaturant gels of normal DNA (N) and DNA from breast carcinomas with p53 mutations. (A) PCR-amplified fragments A of normal DNA and tumor T29 run at 50.7% denaturant. (B) PCR-amplified fragments B of normal DNA and tumor T83 run at 51.7% denaturant. (C) PCR-amplified fragments C of normal DNA and tumor T50 run at 32% denaturant. (D) PCR-amplified fragments D of normal DNA and tumor 119 (T119) run at 50.7% denaturant.

and OsO_4; this should identify 100% of mutations but in practice a small proportion of GT mismatches may be missed.

3) With one fragment labelled and modification by HA only; theoretically 50% of mutations should be found, but in practice, in p53, about 70% were detected by Prosser. This is probably because approximately 70% of p53 mutations in breast cancer occur at GC base pairs.[62]

4) For practical reasons, OsO_4 modification alone is never used since the mutant bands are less pronounced than with HA modification, and it is possible to miss a small proportion of GT mismatches.

An additional benefit of the method is that the mutation is precisely located in one of two positions so that subsequent sequencing is minimal. When directly sequencing PCR products, this is an advantage. A final check that the sequence of the mutation is correct can be made by theoretically considering the outcome of the HA and OsO_4 modification of the mutant fragment.

The technique does not determine homozygosity or heterozygosity of the alleles, but this is achieved in subsequent sequencing. Fragments up to 1kb, and probably greater than 1kb, can be analyzed if the sample is doubled-loaded on the sequencing gel (as is done with sequencing reactions, to space out the region near the top of the fragment). The HOT technique requires no special equipment, can use fragments in the order of 1kb in length and, in addition to detecting the presence or absence of a mutation, actually indicates precisely where the mutation resides. The major drawback of the HOT technique is that it uses hazardous chemicals and is relatively, though not prohibitively, labor-intensive for routine screening.

Each of these three techniques consequently has its merit. For a rapid survey of a set of samples, when what is required is a reasonable estimate of the proportion of samples carrying a mutation in a particular gene, SSCP might be the method of choice, especially as it is easy to set up and run and requires no special equipment. If the aim is to look for one or two mutations in a gene (in order, for instance, to implicate a particular

gene with a particular disease), where 100% detection is important but where there is no intention of screening a large number of samples, HOT might be the method of choice. It is relatively easy to set up, requires no special equipment, can scan in >1kb segments and gives precise information regarding the position of mutations found. When planning to accurately screen a large number of samples, however, the setting up (theoretical and/or experimental) for DGGE/CDGE is worth the effort.

When both techniques are up and running, DGGE/CDGE aims for 100% mutation detection but is much less labor intensive than HOT for screening purposes and uses none of the hazardous chemicals. CDGE avoids the routine pouring of gradient gels but needs careful selection of optimal conditions.

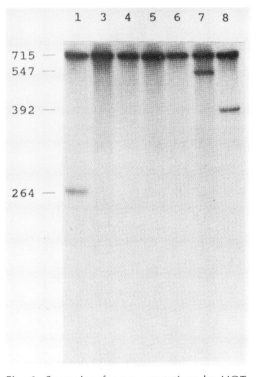

Fig. 6. Screening for new mutations by HOT. Hydroxylamine analysis of exon H in hemophilia B. Lanes 1-7: affected males with unknown mutations. Lane 8: patient London (known mutation in exon H). Reproduced by kind permission of Dr. D. Bentley, Guy's Hospital, London, UK and by permission of Oxford University Press, (Nucleic Acids Res 1989; 17:(9) 3353).

PROTOCOLS FOR SSCP, CDGE AND HOT

THE SSCP TECHNIQUE

Standard PCR is carried out using 32P-dCTP (100 ng DNA, 3 µCi label in a 25 µl reaction). PCR products can be diluted 1/10 in 0.1% SDS and 10mM EDTA but we no longer do this. Eight microliters of 95% formamide stop solution is added to 8µl of PCR reaction, heated to 90°C for 3 min, cooled on ice and loaded onto a 4.5% nondenaturing polyacrylamide gel. The most sensitive running conditions vary with the gene being tested for *ras* these are 1xTBE plus 10% glycerol at room temperature[49] and for p53 are 0.5xTBE without glycerol at 4°C (Table 1).[63]

THE CDGE TECHNIQUE

PCR is carried out using sets of primers to amplify about 300-400bp.[53-55] Theoretical melting profiles of amplified fragments are produced using a computer program.[64] Perpendicular denaturing gradient gels contain 12.5% acrylamide in TAE (0.04 M Tris-acetate, 0.001 M EDTA, pH8) and varying concentrations of denaturant consisting of urea and formamide (100% denaturant corresponds to 7M urea and 40% formamide). Gradient gels are cast with a gravitational gradient maker. The PCR product is loaded along the top of the gel and run perpendicularly to the denaturing gradient. Gels were run submerged in TAE buffer at 55°C to 60°C at constant voltage (80V) for approximately 2-4 hours, depending on the size as well as melting behavior of the actual fragment, using a modified Mini Protean Slab Electrophoresis Cell (Bio-Rad Laboratories). The gels used in CDGE contained the same chemicals as for DGGE, but with a uniform concentration of denaturant through the gel. Running conditions were as for DGGE. After electrophoresis, the gels are stained in ethidium bromide and photographed using a UV transilluminator (Figs. 4 and 5).

THE HOT TECHNIQUE

Standard 100 µl PCR reactions are carried out using oligos which can PCR up to 1kb. The products are excised from TAE/low melting agarose gels (BRL) and Genecleaned (Stratech Scientific). PCR amplified wild-type DNA is kinase-labelled with ^{32}PATP (100 ng DNA, 10 µCi ^{32}p-dATP, 8 units of T4 polynucleotide kinase in 20 µl volume of 1 x kinase buffer). 10 ng of kinase labelled wild-type DNA is added to 100 ng of cold potential mutant DNA in 0.3 M NaCl 0.1 M Tris, pH8, 10 µl volume, overlayered with paraffin oil, heated to 100°C for 5 min, then incubated at 65°C overnight. The heteroduplex is cleaned and precipitated with 15 µg of mussel glycogen (BCL) and taken up in 13 µl of TE. The 13 µl volume is divided into two tubes: 7 µl for hydroxylamine hydrochloride (HA) (Aldrich) modification and 6 µl for osmium tetroxide (OsO_4) (Aldrich or Johnson Matthey) modification.

For HA modification, 20 µl of a fresh 4M solution of HA titrated to pH6 with diethylamide (Aldrich) is added to the 7 µl of heteroduplex and incubated at 37°C for 2 hr. For OsO_4 modification, 2.5 µl of 10 x OsO_4 buffer (100 mM Tris pH 7.7, 10 mM EDTA, 15% pyridine) is added to the 6 µl of heteroduplex, then 15 µl of freshly diluted OsO_4 (1/10 of 8% solution). Incubation is at 37°C for 10 min. The reactions in each modification are stopped with 200 µl 0.3 M Na acetate, then ethanol precipitated after the addition of 5 µl of dextran (MW 2 x 10^6, 20 mg/ml). To the lightly dried pellet in each tube is added 50 µl of freshly diluted 1 M solution of piperidine (Aldrich). The tubes were vortexed and left for 30 min at 90°C. The reaction was stopped and precipitated as before, adding no further dextran. The pellet is taken up in 5 µl TE, 2 µl formamide loading dye and run on a 6% acrylamide/urea gel at room temperature until the bromophenol blue dye reachs the bottom (less than 2 hrs). For longer fragments (800bp ->1kb), the first loading would be run for 4-6 hr and a second loading the usual 2 hr.

^{35}S has been used for HOT[65] but we would not recommend it for SSCP.

Once a mutation is detected, it can be confirmed by direct sequencing (Chapter 6: Sequencing PCR Products) or allele-specific hybridization. The latter uses a labelled mutant probe which is hybridized to Southern blotted PCR product containing the mutant.

Table 1. Mutations characterized by 3 techniques: SSCP, CDGE and HOT

Method of Detection

Sample Number	p53 Codon	Exon	Mutation	SSCP 0.5x TBE* 4°C	1x TBE 4°C	0.5x TBE RT 5% Gly*	Overall Result	CDGE	HOT
A1	151	5	cCc>cAc					$-_1$	M
A2	152	5	cCg>cTg					$-_1$	M
A3	−	−	−					−	?
A4	−	−	−					−	−
A5	175	5	cGc>cAc					M	M
A8	176	5	tGc>tTc					M	M
A9	−	−	−					−	?
A10	238	7	tGt>tTt					M	M
A11	238	7	tgT>tgA					ND	M
A12	245	7	Ggc>Agc					M	M
A13	248	7	Ctt>Tgg					M	M
A15	248	7	cGg>cAg					M	M
								6/8	9/9
B1	135	5	tGc>tTc	M	M	M	M		M
B2	−	−	−	−	−	−	−		−
B3	−	−	−	−	−	−	−		−
B4	−	−	−	−	−	−	−		−
B5	175	5	cGc>cAc	M	−	?	M		M
B6	204	6	Gag>Tag	−	−	?	−		M
B7	239	7	aAc>aCc	M	M	M	M		M
B8	181	5	cGc>cAc	M	−	−	M		M
B9	248	7	Cgg>Tgg	M	M	M	M		M
B10	238	7	tGt>tTt	M	M	M	M		M
B11$_3$	−	−	−	?	−	?	−		−
B12	134	5	Ttt>Ctt	M	M	M	M		M
BD19	−	−	−	−	−	−	−		−
BD20	282	8	Cgg>Tgg	M	−	M	M		M
BD31	276	8	gCc>gAc	M	−	M	M		M
							9/10		10/10
C1	179	5	Cat>Tat	−	−	−	−	M	
C2	182	5	tgC>tgA	M	−	M	M	M	
C3	178	5	cAc>cCc	M	−	M	M	M	
C4	−	−	−	−	−	−	−	−	
C5	175	5	cGc>cAc	M	M	M	M	M	
C6	163	6	Tac>Aac	M	?	−	M	M	
C7	285	8	Gag>Aag	M	M	−	M	ND	
C8	276	8	Gcc>Ccc	M	M	M	M	M	
C9	−	−	−	−	−	−	−	−	
C10	273	8	cGt>cAt	M	M	−	M	M	
C11	273	8	cGg>cTt	M	−	−	M	M	
C12	282	8	cGg>cTg	M	M	−	M	M	
							9/10	9/9	

— normal ? equivocal scored as normal
M mutant
ND not done

	Total	18/20	15/17	19/19
		90%	88%	100%

(100% with** altered conditions)

*1 x TBE = 0.089M Tris-borate
   0.089M Boric Acid
   0.002M EDTA

Gly = glycerol
**After breaking the code in the blinded study, the conditions
could be altered to attain 100% sensitivity
from Condie et al[63]

The latter has been used to detect the different forms of mutant *ras*.[66] Stringent washing in 3M tetramethyl ammonium chloride removes nonspecifically bound probe.[67]

TRANSLOCATIONS AND RESIDUAL DISEASE

PCR can be used for diagnosis or the detection of minimal residual disease. Certain chromosomal translocations occur commonly in follicular non-Hodgkin's lymphomas (t14;18) and chronic myeloid leukemia (Philadelphia chromosome positive (t9;22). In follicular non-Hodgkin's lymphoma, the translocation involves the immunoglobin heavy chain region on chromosome 14 and the BCL2 region (a putative oncogene) on chromosome 18. The primers used hybridize to the regions flanking the translocation and will therefore only amplify the intervening DNA when the translocation is present. If the translocation is not present, the primers anneal to different chromosomes and PCR is impossible. The translocation is found in the majority of follicular non-Hodgkin's lymphomas and is often present even if not cytogenetically detectable.[68] In over half the cases, the break point on chromosome 18 is within a region covering 150 base pairs called the major break point region or break point cluster region (BCR). The sensitivity of PCR allows detection of a malignant cell within a million normal ones and residual disease can be detected even in the presence of a histologically negative marrow.[69,70] The sensitivity is greater than 10,000-fold improvement over Southern blotting. Examples of protocols for amplifying across the translocation can be found in Crescenzi et al 1988.[71]

In the case of chronic myeloid leukemia, the break point on chromosome 9 can occur over a very large area and so potentially large fragments need to be amplified which is not always possible by PCR. This problem can be overcome by basing the amplification on messenger RNA rather than genomic DNA. This is performed on cDNA synthesized from messenger RNA by reverse transcriptase.[72]

GENE EXPRESSION

Gene expression can be detected by several methods: the observation of biochemical or phenotypic changes in cells, in situ hybridization,[73] Northern blotting,[74] S1 nuclease assays[75] or RNA protection studies.[76] RT or exon-connection PCR is based upon the PCR across exons in cDNA made by reverse transcriptase conversion from RNA, using either a random hexamers or specific primers.

RNA is isolated and it is best to use purified RNA although even unpurified cytoplasmic RNA can give good results. About 1µg of total cytoplasmic RNA is sufficient for the amplification of rare messenger RNA species where there may be only one or a few copies per cell. In such an amount about 50-100,000 molecules or more of RNA should be present. The isolation of poly A enriched RNA is therefore not necessary. The same PCR buffer can be used for both reverse transcriptase and PCR reactions. The choice of reverse transcriptase is not crucial; a suitable one would be Mo-MuLV reverse transcriptase from Bethesda Research Labs and Cetus Taq polymerase is used in the PCR reaction. For first strand cDNA synthesis the primers used can either be random hexamers, oligo-dT or the downstream specific PCR primer. If random hexamers or oligo-dT are used, about 0.1 µg per reaction is needed. If using a downstream primer about 10-50 pmoles is optimal. Removal of the RNA template is not necessary because the 90° C heat treatment inactivates the reverse transcriptase and denatures the RNA-DNA hybrids. The remaining RNA template does not seem to interfere with the PCR. It is best to titrate the primer oligonucleotides to find the lowest amount to give a good amplified product, since this gives the cleanest and the most efficient amplification. This should be optimized before each sequence is amplified. The primers should be selected so that they reside in separate exons to enable one to detect by looking at the size of product if there is any contaminating DNA which has been amplified. If the genomic structure is not known, primers separated by 300–400 bases in the 5ʹ coding region of the gene can be used since exons greater than this size are rare in vertebrates[77] and the primers will then probably anneal to different exons. If the gene has no introns, then DNAase treatment of the RNA is necessary to remove the DNA.

The following is a suggested protocol for RT-PCR.[78]

METHODS

REAGENTS FOR RT-PCR REACTIONS

1. 10X PCR buffer: 500 mM KCl, 200 mM Tris.HCl (pH 8.4 at room temp), 25 mM $MgCl_2$ and 1 mg/ml nuclease-free BSA. This solution is made by combining RNase-free, autoclaved stock solutions. The BSA is not autoclaved but added to 1 mg/ml from a 10 mg/ml stock. The nuclease-free BSA may be obtained from Bethesda Research Labs (BRL).

2. Deoxynucleotide Triphosphates: Neutralized, 100 mM solutions purchased from Pharmacia. The dNTPs are combined to make a 10 mM stock solution of each dNTP using 10 mM Tris.HCl (pH 7.5) as diluent.

3. RNasin: Purchased from Promega Corporation at 20-40 units/μl.

4. Random hexamers: 100 pm/μl solution in TE (10 mM Tris.HCl, 1 mM EDTA, pH 8.0). The hexamers are synthesized in-house or can be obtained from Pharmacia.

5. PCR Primers: The primers are usually 18-22 bases in length and dissolved in TE at 10-100 pm/μl. (see Chapter 3: Designing Oligos).

6. Reverse transcriptase: Mo-MuLV obtained from Bethesda Research Labs at 200 units/μl. Other sources of enzymes are suitable.

7. Taq polymerase: Obtained from Perkin Elmer/Cetus at 5 units/μl.

8. Light white mineral oil: from SIGMA.

9. Chloroform: Any reagent grade and saturated with TE.

10. Microfuge tubes.

11. NuSieve and ME agarose: Obtained from FMC.

12. TEA electrophoresis buffer: 40 mM Tris.HCl, 1 mM EDTA and 5 mM sodium acetate, pH 7.5

13. DNA Marker: from BRL.

Note: Use autoclaved tubes and solutions wherever possible and wear gloves to prevent nuclease contamination from fingers.

REVERSE TRANSCRIPTASE REACTION

In a final volume of 20 μl 1X PCR buffer assemble the following: 1mM of each dNTP, 1 unit/μl RNasin, 100 pmoles of random hexamer, 1 μg of total or cytoplasmic RNA and 100-200 units of BRL Mo-MuLV reverse transcriptase. Incubate 10 min at room temp, then 30-60 min at 42° C. To stop reaction, heat tube in 95° C water bath for 5-10 min, then quick chill on ice. (NB—other manufacturers of reverse transcriptase recommend different amounts of enzyme than BRL). It is helpful to heat treat the RNA sample at 90° C for 5 min and quick chill before adding an aliquot to the reaction mix. The heat treatment breaks up RNA aggregates and some secondary structure which may inhibit the priming step.

PCR REACTION

To the heat treated 20 μl reverse transcriptase reaction add 80 μl of 1X PCR buffer containing 10-50 pmoles each of upstream and downstream primer and 1-2 units of Taq polymerase. Place 1 drop (about 50 μl of the mineral oil on top of the solution to prevent evaporation). Set up PCR cycles as outlined in Chapter 2.

One-step RT-PCR reactions

RT-PCR can now be done in one step. It involves the use of heat stable reverse transcriptase.[79]

The detection of gene expression

Several reports show the presence of messenger RNA can be demonstrated using RT-PCR in from 10 to 1000 cells.[80,81] This technology can also analyze messenger RNA from hundreds of colonies of cells to look for the expression of genes for growth or differentiation factors, where antibody detection methods would not be possible. This method can also be used in transgenic animals to know whether the transgene is expressed[82] and also its site-specific expression. Since the method is very sensitive, the animal does not have to be sacrificed, and tests can be done on small amounts of material.

Detection of RNA sequences has been used in diagnosis; for example, in chronic myeloid leukemia where there is a (t9:22) translocation and the break point can occur over a wide area precluding PCR from genomic DNA. In promyelocytic leukemia, a fusion gene is formed by a chromosomal translocation and its product can be detected by RT-PCR of the mRNA.[83] RT-PCR is also being used to detect expression of the multiple drug resistance genes.[84]

GENETIC DISORDERS

PCR has revolutionized both the study of genetic disorders and their diagnosis in the diagnostic laboratory. The advantage of PCR is that it only needs small amounts of DNA and it is very rapid. For example before PCR was introduced in 1987 to diagnose sickle cell anemia, Southern blotting was used which took two or more weeks from the date of antenatal sampling.

GENETIC LINKAGE STUDIES

Genetic linkage is the cosegregation of a gene or DNA marker with a disease (Fig. 7). The closer the marker is to the disease gene, the less likely that recombination will occur during meiosis, and the tighter the linkage. The original markers used were markers such as blood groups. Restriction fragment length polymorphisms (RFLPs) were then used as markers, but in general they are not as highly polymorphic as the new repeat markers and so the percentage of individuals that are informative (heterozygotes) is less. Introduction of PCR-based repeat markers has not only speeded up linkage analysis but also made it possible to work on very small amounts of DNA from families. Since most PCR-based repeat markers are more polymorphic, these markers are also more informative. Although some VNTR (variable number of tandem repeats) markers can be subjected to PCR, PCR is usually performed on tri- or dinucleotide repeats (CA repeats or microsatellite markers). The use of such markers has, for example, narrowed down the region within which lies one of the genes for breast and ovarian cancer (BrCa1 on chromosome 17q).[85,86] Such PCR based CA repeat markers have now been isolated over the whole human genome and are available through Genethon.[87] Since the alleles amplified by PCR might differ by only two or three base pairs,

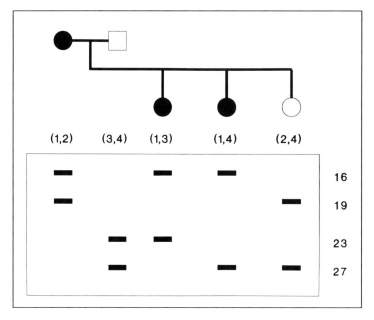

Fig. 7. Demonstration of pattern of inheritance of PCR-based repeat markers in a family with affected members (filled in symbols). The allele 16 cosegregates with the disease.
Symbols: square: male   circle: female

the products are run on a sequencing polyacrylamide gel to improve resolution. They can either be visualised autoradiographically by labelling one of the primers with ^{32}P ATP or by fluorescent visualization of the nonradioactive product after blotting.

GENETIC DIAGNOSIS USING PCR

Genetic diagnosis can either be made by studying the gene directly to look for mutations or by doing linkage analysis similar to that used for finding new disease genes as described above. The latter requires that samples are available (either from blood, or from paraffin embedded tissue from individuals that may be deceased) to enable the pattern of markers flanking the disease gene to be mapped through the family.

THE USE OF PCR IN THE DETECTION OF POINT MUTATIONS FOR DIAGNOSIS

Three techniques that are used following PCR in the diagnosis of point mutations are:
    —restriction enzyme analysis[88]
    —dot blot hybridization[89]
    —direct sequencing[90]
In the antenatal diagnosis of sickle cell anemia, fetal cells obtained at chorionic villous sampling or amniocentesis are boiled in salt and alkali and PCR of a 725 base pair product is carried out. This is digested with the enzyme Cvn1 and run on an agarose gel which is then stained with ethidium bromide. The primers have been chosen so that the product contains 2 Cvn1 sites and therefore three constant bands are seen in the normal samples. In patients with a sickle cell betaglobin chain, two of these sites are lost and so no digestion occurs.

Phenylketonuria is an autosomal recessive disease caused by an inborn error of metabolism. It results from the absence of hepatic phenylalanine hydroxylase resulting in an accumulation of phenylalanine which causes mental retardation.[91] There are about 12 RFLP types at the phenylalanine hydroxylase locus in the Northern European population. About 90% of the alleles are confined to types 1-4. About 40% of mutants occur at the intron-exon 12 boundary and consist

of a single base substitution and about another 20% of mutations are caused by a substitution at a specific codon in exon 12. Oligonucleotide probes specific for these mutations can detect phenylketonuria carriers. The use of PCR in small amounts of material such as a small needle prick blood sample can be PCR'd for a dot blot hybridization analysis. No hybridization is seen with a mutant allele probe in normal controls.

X-linked Duchenne muscular dystrophy (DMD) effects 1 in 3500 male births and causes fatal muscular weakness in the teenage years. Partial intragenic deletions in the coding region of the gene account for more than 50% of the mutations which cause the disease.[92] Analysis of these mutations has been revolutionized by the use of multiple primers in a single PCR. A DMD gene is greater than two megabases and contains at least 70 exons separated by an average intron size of 35kb.[92] Although the deletions that give rise to the disease occur heterogeneously they are usually large, encompassing several hundred kilobases, and are concentrated around two areas of the gene. Multiplex PCR amplifies nine separate regions of the DMD gene in one PCR reaction. This identifies 80-90% of all DMD gene deletions.[93,94]

The p53 gene is the most common oncogene altered in human cancers.[95] Although most mutations are somatic, it can also be mutated in the germline and about 30 such mutations have been described to date. Most of them cause either the Li-Fraumeni or the Li-Fraumeni-like syndrome. The Li-Fraumeni syndrome was defined by Li and Fraumeni in 1969 and 1988 as the association of sarcoma at less than 45 years in the proband of a family associated with either early onset breast cancer or brain tumor, leukemia, adrenocortical carcinoma or sarcoma in a first degree relative and a tumor at less than 45 years or sarcoma at any age in another close relative.[96] The Li-Fraumeni-like syndrome has been described by some workers as consisting of two close relatives each with different tumors at <60 years and the tumors are those seen in the Li-Fraumeni syndrome. About half of Li-Fraumeni families may be due to mutations in the germline in the p53 gene (Li, personal communication); the exact proportion of Li-Fraumeni-like families due to such mutations is unknown but is probably less than 50%. Follow-up of carriers in Li-Fraumeni kindreds has shown that they are at increased risk of early onset cancer before the age of 45; in particular, female carriers are at 18 times the population risk of early onset breast cancer.[97] The p53 gene consists of 11 exons, coding for a 393 amino acid product. The gene contains five conserved regions within which most of the mutations are located. Most of the germline mutations are missense mutations, but a few are small insertions or deletions of a few base pairs.[98,99] Predictive testing for germline mutations in the p53 gene is in its infancy[100] but will almost certainly be offered routinely to Li-Fraumeni families and maybe also families with a Li-Fraumeni-like pattern in the future.[101] Such testing will either involve a general screening technique to screen the whole gene, followed by sequencing of any abnormal areas or sequencing of the full length transcript.

SINGLE CELL PCR AND ANTENATAL DIAGNOSIS

Theoretically, PCR should be able to amplify a single copy of DNA and therefore is applicable to small samples and even single cells. Li and Gyllensten[102] have described PCR of single sperm or single diploid cells obtained by micromanipulation or flow sorting. The cells or sperm are lysed in proteinase K and STS (or in the case of sperm, DTT) and after heat inactivation, the PCR reagents are added. Monk at University College, London has used a buccal cell scrape to provide single cells which are then placed between two marker oil droplets in a glass tube. The cell is suspended in the PCR reaction mixture which also contains albumin to reduce the adherence of cells to the side of the tube. PCR then is performed in the tube.

PCR of very small amounts of material is ideal for antenatal testing where the sample size provided is small.[103-6] Some times even inadequate blood samples or chorionic villous samples are submitted for diagnostic DNA studies. Balnaves has described a modification of DNA extraction and PCR which involves the suspension of the sample in a ligase buffer and incubation with proteinase K and the lysates are used in the PCR.[107]

PCR FROM ARCHIVAL MATERIAL AND APPLICATIONS IN FORENSIC SCIENCE

The ability to perform PCR on paraffin-embedded tissues has opened up large areas of research and diagnosis. It is now possible to perform linkage analysis on families even where many individuals are deceased, provided their paraffin-embedded tumor material is available. Large archives of paraffin-embedded material are available for study whereas frozen tissue banks are often less extensive.

EXTRACTION OF DNA
FROM PARAFFIN-EMBEDDED TISSUES

DNA up to about 300-400 base pairs in length can be extracted from paraffin-embedded specimens either using a phenol chloroform extraction technique or lysis buffer and proteinase K alone; PCR can then be performed on the supernatant.[108-112] We have found that the use of bovine serum albumin in the PCR at a final concentration of 100 µg/ml improves the PCR yield. The use of a hot start further improves this, but not dramatically so (Chapter 3: Problems and Pitfalls). DNA extracted from paraffin-embedded sections contains inhibitors which can be shown to be present because they prevent the PCR of DNA from fresh blood. Since DNA in paraffin-embedded sections is extensively degraded, amplification of greater than 400 base pairs is often impossible. The ability to perform PCR of such DNA is related to the method of fixation and the age of the sample. RNA can also be extracted from paraffin-embedded tissue and may be up to a 100 to 200 base pairs long.[113]

Traditionally loss-of-heterozygosity studies to ascertain the sites of tumor suppressor genes have been performed in frozen tissue, comparing the allele pattern in the tumor with that of the DNA from fresh blood. The use of PCR of CA repeat markers in paraffin-embedded tumor specimens which have been microdissected enlarges the number of tumors on which this technique can be performed. The allele status of normal DNA can be ascertained by PCR of normal areas, again obtained by microdissection.[114]

The PCR has been modified by Yap et al[115] so that it can be performed directly on tissue on slides. The PCR reaction is placed on the tissue on the slides and covered with a cover slip. PCR is then performed by laying the slide on the PCR heating block and collecting the PCR mix with a pipette after cycling.

Amplification of ancient DNA

DNA is still present in ancient specimens but is heavily modified.[116] PCR can therefore be used to isolate the DNA from the few copies of intact DNA still present in an ancient sample where most of the DNA will be degraded. Contamination can therefore be a large source of error and in handling specimens plastic gloves should be worn and great attention should be paid to avoiding contamination. DNA samples should be taken from as deep within the tissue as possible since these areas have been less exposed to contaminants. Efficiency of the amplification is greater if smaller fragments are amplified because of the degraded nature of the old DNA. In ancient samples the maximum length of mitochondrial DNA that has been amplified is between 100 and 200 base pairs.[117]

Forensic science

The advantage of PCR in forensic science is its ability to amplify DNA from very small samples such as blood spots or a human hair.[118,119] PCR genetic finger printing can be carried out by performing PCR of the VNTRs (variable number of tandem repeats or minisatellites) sequences of about 30 bases which are repeated end-to-end at intervals in the introns of the genome. The number of times these sets of bases are repeated vary between individuals. Up to six mini-satellites can be amplified in a single PCR.[120] The VNTR markers can be used for paternity testing and individual identification for forensic purposes. The usual techniques for determining genotypes using VNTR loci involve the Southern blot assay which takes several days, use radioactivity and need larger amounts (several micrograms) of DNA than PCR.

References

1. Harper MH, Marselle LM, Gallo RC, Wong-Staal F. Detection of lymphocytes expressing human T-lymphotrophic virus type III in lymph nodes and peripheral blood from infected individuals by in situ hybridization. Proc Natl Acad Sci 1986; 83:772.

2. Steinhauer DA, Holland JJ. Rapid evolution of RNA viruses. Ann Rev Microbiol 1987; 41:409-433.

3. Barre-Sinoussi F, Chermann JC, Rey F, Nugeyre MT, Chamaret S, Gruest J, Dauguet C, Axler-Blin C, Vezinet-Brun F, Rouzioux C, Rezenbaum W, Montagnier L. Science, 1983; 220:868-870.

4. Popovic M, Sarngadharan MG, Read E, Gallo RC. Detection, isolation and continuous production of cytopathic retroviruses (HTLV-III) from patients with AIDS and pre-AIDS. Science, 1984; 224:497-500.

5. Guyader M, Emerman M, Sonigo P, Clavel F, Montagnier L, Alizon M. Genome organization and transactivation of the human immunodeficiency virus type 2. Nature, 1987; 326:662-669.

6. Imagawa DT, Lee MH, Wolinsky SM et al. Human immunodeficiency virus type I infection in homosexual men who remain seronegative for prolonged periods. N Engl J Med, 1989; 320:1458-1462.

7. Rogers MF, Ou Cy, Rayfield M et al. Use of the polymerase chain reaction for early detection of the proviral sequences of human immunodeficiency virus in infants born to seropositive mothers. New York City Collaborative Study of Maternal HIV Transmission and Montefiore Medical Center HIV Perinatal Transmission Study Group. N Engl J Med 1989; 320:1649-1654.

8. Ehrlich G, B Poiesz. Clinical and molecular parameters of HTLV-I infection. Clin Lab Med 1988; 8:65.

9. Bhagavati S, Ehrlich G, Kula R, Kwok S, Sninsky J, Udani V, Poiesz B. Detection of human T-cell lymphoma/leukemia virustype I (HTLV-1) in the spinal fluid and blood of cases of chronic progressive myelopathy and a clinical, radiological and electrophysiological profile of HTLV-1 associated myelopathy. N Engl J Med 1988; 318:1141.

10. Kwok S, Ehrlich G, Poiesz B, Bhagavati, Sninsky J. Characterization of a HTLV-1 sequence from a patient with chronic progressive myelopathy. J Infect Dis 1988; 158:1193.

11. Kwok S, Ehrlich G, Poiesz B, Kalish R and Sninsky J. Enzymatic amplification of HTLV-I viral sequences from peripheral blood monoclear cells and infected tissues. Blood 1988b; 72:1117.

12. Aono Y, Imai J, Tominaga K, Orita S, Sato A, Igarashi H. Rapid, sensitive, specific, quantitative detection of human T-cell improved polymerase chain reaction method with nested primers. Virus Genes 1992; 6:(2) 159-71.

13. Greenberg S, Ehrlich G, Abbott M, Hurwitz A, Waldmann T, Poiesz B. Detection of sequences homologous to human retroviral DNA in multiple sclerosis by gene amplification. Proc Natl Acad Sci USA, 1989; 86:2878-2882.

14. Ehrlich H. PCR protocols: A guide to methods and applications. Innis MA, Gelfand D, Sninsky JH, White T (eds.) Academic Press Inc 1990; 327-8.

15. Kwok S, Kellogg D, Ehrlich G, Poiesz B, Bhagavati S, Sninsky JJ. J Infec Dis 1989; 158:1193-1197.

16. Duggan DB, Ehrlich GD, Davey FP, Kwok S, Sninsky J, Goldberg J, Baltrucki L, Poiesz B. HTLV-1 induced lymphoma mimicking Hodgkin's disease, diagnosis by PCR amplification of specific HTLV-1 sequences in tumor DNA. Blood 1988; 71:1027-1032.

17. Kalyanaraman VS, Sarngadharan MG, Robert-Guroff M, Miyoshi I, Blayney D, Golde D, Gallo RC. A new sub type of human T-cell leukemia virus (HTLV-II) associated with a T-cell variant of hair cell leukemia. Science, 1982; 218:571.

18. Rosenblatt JD, Golde DW, Wachsman W, Giorgi JV, Jacobs A, Schmidt GM, Quan S, Gasson JC, Chen I. N Engl J Med 1986; 315:72.

19. Lee H, Swanson P, Shorty VS, Zack J, Rosenblatt JD, Chen ISY. High rate of HTLV-II infection in seropositive v drug abuses in New Orleans. Science 1989; 244:471-475.

20. Shaw GM, Hahn BH, Arya SK, Groopman JE, Gallo RC, Wong Staal F. Molecular characterization of human T-cell leukemia (lymphotropic) virus type III in the acquired immune deficiency syndrome. Science, 1984; 226:1165-1171.

21. Saag MS, Hahn BH, Gibbons Y, Li Y, Parks ES, Parks WP, Shaw GM. Extensive variation of human immunodeficiency virus type in vitro. Nature (London) 1988; 334:440-444.

22. Goodenow MT, Huet T, Saurin W, Kwok S, Sninsky J, Wain Hobson S. HIV-I isolates are rapidly evolving quasispecies: Evidence for viral mixtures and preferred nucleotide substitutions. JAIDS 1989; 2:344-352.

23. Ou CY, Kwok S, Mitchell SW, Mack DH, Sninsky JJ, Krebs JW, Feorino P, Warfield D, Schochetman G. DNA amplification for direct detection of HIV-I in DNA of peripheral blood mononuclear cells. Science, 1988; 238:295-297.

24. Kwok S, Mack DH, Sninsky JJ, Ehrlich GD, Poiesz BJ, Dock NL, Alter HJ, Mildvan D, Grieco MH. Diagnosis of human immunodeficiency virus in seropositive individuals: Enzymatic amplification of HIV viral sequences in peripheral blood mononuclear cells. In HIV detection by genetic engineering methods (ed P A Luciw, K S Steimer) Marcel Dekker Inc New York:1989.

25. Kellogg DE, Kwok S. PCR Protocols: A guide to methods and applications. Innis MA, Gelfand D, Sninsky J, White T (eds.) Academic Press Inc 1990:339.

26. Laure F, Courgnaud V, Rouzioux C, Blanche S, Veber F, Burgard M, Jacomet C, Griscelli C, Brechot. Detection of HIV I DNA in infants and children by means of the PCR. Lancet 1988; 2:538-541.

27. DeRossi A, Amdori A, Chieco-Bianchi L et al. DeRossi Lancet 1988; 2:278.

28. Weintrub PS, Ulrich PP, JR Edwards, Boucher F, Levy JA, Cowan MJ, Vyas GN. Use of polymerase chain reaction for the early detection of HIV infection in the infants of HIV seropositive women. Aids 1991; 5(7):881-4.

29. Bush CE, Donovan RM, Peterson WR, Jennings MB, Bolton V, Sherman DG, Vanden-Brink KM, Beninsig LA, Godsey JH. Detection of human immunodeficiency virus type 1 RNA in plasma samples from high risk pediatric patients by using the self sustained sequence replication reaction. J Clin Microbiol 1992; 30(2):281-6.

30. Mack DH, Sninsky JJ. A sensitive method for teh identification of uncharacterized viruses related to known virus groups: Hepadnavirus model system. Proc Nat Acad Sci USA 1988; 85:6977-6981.

31. Manzin A, Salvoni G, Bagnarelli P, Menzo S, Carloni G, Clementi M. J Virol Methods. A single step DNA extraction procedure for the detection of serum hepatitus B virus sequences by the polymerase chain reaction. J Virol Methods 1991; 32:2-3 245-53.

32. Shibata D. PCR Protocols A Guide to Methods and Applictions:369.

33. Demmler GJ, Buffone GJ, Schimbor SM, May RA. Detection of cytomegalovirus in urine from newborns by using polymerase chain reaction DNA amplification. J Infec Dis 1988; 158:1177-1184.

34. Shibata D, Martin WJ, Appleman MD et al. J Infec Dis 1988; 158:1185-1192.

35. Zipeto D, Revello MG, Silini E, Parea M, Percivalle E, Zavattoni M, Milanesi G, Gerna G. Development, clinical significance of a diagnostic assay based on the blood samples from immunocompromised patients. J Clin Microbiol 1992; 30:2 527-30.

36. Pfister H. Human papillomaviruses and genital cancer. Adv Cancer Res 1987; 48:113-147.

37. Stanley M. Genital papillomaviruses, polymerase chain reaction and cervical cancer. Genitourin Med 1990; 66:6:415-7.

38. Sharma BK, Luthra UK, Shah KV. Identification of human papillomaviruses in paraffin embedded cervical pathological tissues form Indian women by polymerase chain reaction. Ann Biol Clin Paris 1991; 49:2 93-7.

39. Weiss LM, Movahed LA, Billingham ME, Cleary ML. Detection of Coxsackievirus B3 RNA in myocardial tissues by the polymerase chain reaction. Am J Pathol 1991; 138:2 497-503.

40. Rotbart HA. Human enterovirus infections - molecular approaches to diagnosis and pathogenesis. In Molecular aspects of Picornavirus infection and detection (ed B Semler, E Ehrenfeld) Am Soc of Mircobiol Washington DC:1989; 243-264.

41. Rotbart HA. PCR Protocols A Guide to Methods and Applications:373.

42. Lukenheimer K, Hufert FT, Schmitz H. Detection of Lassa virus RNA in specimens from patients with Lassa fever by using the polymerase chain reaction. J Clin Microbiol 1990; 28:(12), 2689-92.

43. Allard, A, Girones R, Juto P, Wadell G. Polymerase chain reaction for detection of adenoviruses in stool samples. J Clin Microbiol 1990; 28:(12) 2659-67.

44. Lazizi Y, Elfassi E, Pillot J. Detection of hepatitis C virus sequences in sera with controversial serology by nested polymerase chain reaction. J Clin Microbiol, 1992; 30:(4) 931-4.

45. Cotton RGH. Detection of single base changes in nucleic acids. Biochem J 1989; 263, 1-10.

46. Rossiter BJF, Caskey CT. Molecular scanning methods of mutation detection. J Biol Chem 1990; 265:12753-12756.

47. Theophilus BDM, Latham T, Grabowski GA, Smith FI. Comparison of RNase A, chemical cleavage and GC-clamped denaturing gel electrophoresis for the detection of mutations in exon 9 of the human acid b-glucosidase gene. Nucleic Acids Res 1989; 17:7707-7722.

48. Orita M, Iwahana H, Kanazawa H, Hayashi K, Sekiya T. Detection of polymorphisms of human DNA by gel electrophoresis as single-strand conformation polymorphisms. Proc Natl Acad Sci USA 1989a; 86:2766-2770.

49. Orita M, Suzuki Y, Sekiya T, Hayashi K. Rapid and sensitive detection of point mutations and DNA polymorphisms using the polymerase chain reaction. Genomics 1989b; 5:874-879.

50. Spinardi L, Mazars R, Theillet C. Protocols for an improved detection of point mutations by SSCP. Nucleic Acids Res 1991; 19:4009.

51. Fischer SG, Lerman LS. DNA fragments differing by single base-pair substitutions are separated in denaturing gradient gels. Correspondence with melting theory. Proc Natl Acad Sci USA 1983; 80:1579-1583.

52. Myers RM, Maniatis T, Lerman LS. Detection and localization of single base changes by denaturing gradient gel electrophoresis. Methods in Enzymology 1987; 155:501-527.

53. Borresen, A-L, Hovig, E, Smith-Sorensen, B, Malkin, D, Lystad, S, Andersen, IT, Nesland, JM, Isselbacher, KJ, Friend, SH. Constant denaturant gel electrophoresis as a rapid screening technique for p53 mutations. Proc Natl Acad Sci USA 1991a; 88:8405-8409.

54. Borresen A-L, Hovig E, Smith-Sorensen B Lystad S, Brogger A. Advances in Mol Gen 1991b; 4:63-71.

55. Hovig E, Smith-Sorensen B, Brogger A, Borresen A-L. Constant denaturant gel electrophoresis, a modification of denaturing gradient gel electrophoresis, in mutation detection. Mut Res 1991; 262:63-71.

56. Cotton RGH, Rodrigues NR, Campbell RD. Reactivity of cytosine and thymine in single-base-pair mismatches with hydroxylamine and osmium tetroxide and its application to the study of mutations. Proc Natl Acad Sci USA 1988; 85:4397-4401.

57. Montandon AJ, Green PM, Giannelli F, Bentley DR. Direct detection of point mutations by mismatch analysis: Application to hemophilia B. Nucleic Acids Res 1989; 17:9.

58. Prosser J, Elder PA, Condie A, Macfadyen I, Steel CM, Evans HJ. Mutations in p53 do not account for heritable breast cancer: A study in five affected families. Br J Cancer 1991; 63:181-184.

59. Orita M, Sekiya T, Hayashi K. DNA sequence polymorphisms in Alu repeats. Genomics 1990; 8:271-278.

60. Savov A, Angelicheva D, Jordanova A, Eigel A, Kalaydjiera L. High percentage acrylamide gels improve resolution in SSCP analysis. Nuc Acids Res 1992; 20:(24) 6741-2.

61. Ainsworth PJ, Surh LC, Coulter-Mackie MB. Diagnostic single strand conformational polymorphism (SSCP): A simplified non-radioisotopic method as applied to a Tay-Sachs B1 variant. Nucleic Acids Res 1991; 19:405.

62. Coles C, Condie A, Chetty U, Steel CM, Evans HJ, Prosser J. p53 mutations in breast cancer. Cancer Res 1992; 52:(19) 5291-8.

63. Condie A, Eeles RA, Borresen A-L, Coles C, Cooper CS, Prosser J. Detection of point mutations in the p53 gene. Comparison of SSCP, CDGE and HOT. Human Mutation (1993) 2, 58-66.

64. Lerman, LS, Silverstein K. Computational simulation of DNA melting and its application to denaturing gradient gel electrophoresis. Methods in Enzymology 1987; 155:482-501.

65. Saleeba JA, Cotton RGH. ^{35}S-labelled probes improve detection of mismatched base pairs by chemical cleavage. Nuc Acids Res 1991; 19(7) 1712.

66. Bos JL. The ras gene family and human carcinogenesis. Mutat Res 1988; 195:255 271.

67. Wood WI, Gitschier J, Lasky LA et al. Base composition-independent hybridization in tetramethylammonium chloride: A method for ologonucleotide screening of highly complex gene libraries. Proc Natl Acad Sci USA 1985; 82:1585-1588.

68. Yunis JJ. The chromosomal basis of human neoplasia. Science 1983; 221:227-236.

69. Negin RS, Blume KG. The use of the polymerase chain reaction for the detection of minimal residual malignant disease. Blood 1991; 78 (2) 255-8.

70. Lee MS, Kantarjian H, Talpaz M, Freireich EJ, Deisseroth A, Trujillo JM, Stass SA. Detection of minimal residual disease by polymerase chain reaction in Philadelphia chromosome positive chronic myelogenous leukemia following interferon therapy. Blood 1992; 79:(8) 1920-3.

71. Crescenzi M. PCR Protocols: A guide to methods and applications. Innis MA, Gelfand D, Sninsky J, White T (eds.) Academic Press Inc 1990:393.

72. Dobrovic A, Trainer KJ, Morley AA. Detection of the molecular abnormality in chronic myeloid leukaemia by use of the polymerase chain reaction. Blood 1988; 27:2063-2065.

73. Angerer RC, Cox KH, Angerer LM. Genetic Engineering 1985; 7:43-65.

74. Thomas PS. Hybridization of denatured RNA and small DNA fragments transferred to nitrocellulose. Proc Natl Acad Sci USA 1980; 77:5201-5205.

75. Berk AJ, Sharp PA. Cell 1977; 12:721-732.

76. Melton DA, Krieg PA, Rabagliati MR, Maniatis T, Zinn K, Green MR. Nucl Acids Res 1984; 7:1175-1193.

77. Hawkins JD. A survey on intro and exon lengths. Nucl Acids Res 1988; 16:9893-9908.

78. Kawasaki et al. PCR Technology, Principles and Applications for DNA Amplification Ed Ehrlich H Stockton Press 1989; 90.

79. Myers TW, Gelfand DH. Reverse transcription and DNA amplification by a Thermus thermophilus. Biochemistry 1991; 6 30, 31:7661-6.

80. Kawasaki ES, Clark SS, Coyne MY, Smith SD, Champlin R, Witte ON, McCormick FP. Diagnosis of chronic myeloid and acute lymphocytic leukemias by detection of leukemia-specific mRNA sequences amplified in vitro. Proc Natl Acad Sci USA 1987; 85:5698-5702.

81. Noonan KE, Roninson IB. mRNA phenotyping by enzymatic amplification of randomly primed cDNA. Nucleic Acids Res 1988; 16:10366.

82. Hanley T, Merlie JP. Transgene detection in unpurified mouse tail DNA by polymerase chain reaction. Biotech. 1991; 10:(1) 56.

83. Chen SJ, Chen Z, Chen A, Tong JH, Dong S, Wang ZY, Waxman S, Zelent A. Occurrence of distinct PML-RAR-alpha fusion gene isoforms in patients with acute promyelocytic leukemia detected by reverse transcriptase/polymerase chain reaction. Oncogene 1992; 7 (6):1223-32.

84. Sugawara I, Watanabe M, Masunaga A, Itoyama S, Ueda K. Primer independent amplification of mdr1 mRNA by polymerase chain reaction. Jpn J Cancer Res 1992; 83:2 131-3.

85. Hall JM, Lee MK, Newman B et al. Linkage analysis of early onset familial breast Cancer to chromosome 17q21. Science 1990; 250:1684-9.

86. Easton DF, Bishop DT, Ford D, Crockford GP, the Breast Cancer Linkage Consortium. Genetic linkage analysis in familial breast and ovarian cancer - results form 214 families. Am J Hum Genet 1993; 52:678-701.

87. Weissenbach J. The Genethon Microsatellite map catalogue. Genethon 1992; Human Genome Research Centre, Evry, France.

88. Saiki RK, Bugawan TL, Horn GT, Mullis KB, Erlich HA. Analysis of enzymatically amplified β-globin and HLA-DQa DNA with allele specific oligonucleotide probes. Nature 1987; 324:163.

89. Kogan SC, Doherty M, Gitschier J. An improved method for prenatal diagnosis of genetic diseases by analysis of amplified DNA sequences. Application to hemophilia. A N Engl J Med 1987; 317 985.

90. Saiki RK, Gelfand DH, Stoffel S, Scharf SJ, Higuchi R, Horne GT, Mullis kB, Erlich HA. Primer directed enzymatic amplification of DNA with a thermostable DNA polymerase. Science 1988; 239 :489.

91. Woo SLC. Molecular basis and population genetics of phenylketonuria. Biochemistry 1989; 28:1.

92. Koenig M, Hoffman EP, Bertelson CJ, Monaco AP, Feener C, Kunkel LM. Complete cloning of the Duchenne muscular dystrophy DMD cDNA and preliminary genomic organisation of the DMD gene in normal and affected individuals. Cell 1987; 50:509-517.

93. Chamberlain JS, Gibbs RA, Ranier JE, Nguyen PN, Caskey CT. Deletion screening of the Duchenne muscular dystrophy locus via multiplex DNA amplification. Nuc Acids Res 1988; 16:11141-11156.

94. Chamberlain JS. PCR Protocols: A guide to methods and applications. Innis MA, Gelfand D, Sninsky J, White T (eds.) Academic Press Inc 1990:33.

95. Hollstein MC, Sidrausky D, Vogelstein R, Harris CC. p53 mutations in human cancers. Science 1991; 253:49-53.

96. Li FP, Fraumeni JF Jr, Mulvihill JJ, Blattner WA, Dreyfus MG, Tucker MA, Miller RW. A cancer family syndrome in 24 kindreds. Cancer Research 1988; 48:5358-62.

97. Garber JE, Goldstein AM, Kantor AF, Dreyfus MG, Fraumeni JF Jr and Li F. Follow up study of 24 families with Li-Fraumeni syndrome. Cancer Research 1991; 51:6094-7.

98. Malkin D, Li FP, Strong LC et al. Germline p53 mutations in a familial syndrome of breast cancer sarcomas and other neoplasms. Science 1990; 250:1233-8.

99. Toguchida J, Yamaguchi T, Dayton SH, et al:Prevalence and spectrum of germline mutations of p53 gene among patients with sarcoma. New England Journal of Medicine 1992; 326:1301-8.

100. Eeles RA. Predictive testing for germline mutations in the p53 gene: Are all the questions answered? Eur J Cancer 1993 in press.

101. Eeles RA,. Warren W, Knee G et al. Constitutional mutation in exon 8 of the p53 gene in a patient with multiple independent primary tumors: Molecular and immunohistochemical findings. Oncogene 1993; 8:1269-76.

102. Li H, Gyllensten UB, Cui X, Saiki FRK, Erlich HA, Arnheim N. Amplification and analysis of DNA sequences in single human sperm and diploid cells. Nature London 1988; 335:414-417.

103. Gasparini P, Novelli G, Savoia A, Dallapiccola B, Pignatti PF. First trimester prenatal diagnosis of cystic fibrosis using the polymerase chain reaction: Report of eight cases. Prenat Diagn 1989; 9:349-355.

104. Kogan SC, Doherty M, Gitschier J. An improved method for prenatal diagnosis of genetic diseases by analysis of amplified DNA sequences. Application to hemophilia A. N Engl J Med 1987; 317:985-990.

105. McIntosh I, Curtis A, Millan FA, Brock DJ. Prenatal exclusion testing for Huntington disease using the polymerase chain reaction. Am J Med Genet 1989; 32:274-276.

106. Newton CR, Kalsheker N, Graham A et al. Diagnosis of alpha 1 antitryspin deficiency by enzymatic amplification of human genomic DNA and direct sequencing of polymerase chain reaction products. Nucleic Acids Res 1988; 16:8233-8243.

107. Balnaves ME, Nasioulas S, Dahl HH, Forrest S. Direct PCR from CVS and blood lysates for detection of cystic fibrosis and Duchenne muscular dystrophy deletions.[published erratum appears in Nucleic Acids Res 1991 11d:19(9):2537]. Nucleic Acids Res 1991; 19(5):1155.

108. Cooper CS, Stratton MR. Extraction and enzymatic amplification of DNA from paraffin embedded specimens. Methods in Molecular Biology, Vol 9 Protocols in Human Molecular Genetics. 1991.

109. Greer CE, Peterson SL, Kiviat NB et al. PCR amplification from paraffin-embedded tissues: Effects of fixative and fixation time. Am J Clin Pathol 1991; 95:117-124.

110. Fey MF, Pilkington SP, Summers C et al. Molecular diagnosis of haematological disorders using DNA from stored bone marrow slides. Br J Haematol 1987; 67:489-492.

111. Shibata DK. Detection of human papillomarvirus in paraffin-embedded tissues using a new in vitro DNA amplification procedure. Am J Clin Pathol 1987; 88:524.

112. Grunewald K, Feichtinger H, Weyerer K, Dietze O, Lyons J. DNA isolated from plastic embedded tissue is suitable for PCR. Nuc Acid Res Vol 1990; 18:20.

113. Stanta G, Schneider C. RNA extracted from paraffin-embedded human tissues is amenable to analysis by PCR amplification. Biotech 1991; 11:3.

114. Bianchi A, Navone MN, Conti CJ. Detection of loss of heterozygosity in formalin fixed paraffin embedded tumor specimens by the polymerase chain reaction. Am J Path 1990; 38.

115. Yap EPH, McGee JOD. Slide PCR: DNA amplification from cell samples on microscopic glass slides. Nuc Acid Res 1991; 19:No. 15.

116. Paabo S. Ancient DNA: Extraction, characterization, molecular cloning and enzymatic amplification. Proc Natl Acad Sci USA 1989; 86:1939-1943.

117. Paabo S, Gifford JA, Wilson AC. Mitochondrial DNA sequences from a 7000 year old brain. Nucl Acids 1988b; 16:9775-9787.

118. Witt M, Erickson RP. A rapid method for detection of Y chromosomal DNA from dried blood specimens by the polymerase chain reaction. Hum Genet 1989; 82:271-274.

119. Higuchi R, von Beroldingen CH, Sensabaugh GF, Erlich HA. DNA typing form single hairs. Nature 1988; 332:543-546.

120. Jeffreys AJ, Wilson V, Neumann R, Keyte J. Amplification of human minisatellies by the polymerase chain reaction: Towards DNA fingerprinting of single cells. Nucl Acids Res 1988; 16:10953-10971.

CHAPTER 6

RESEARCH APPLICATIONS

QUANTITATIVE PCR

QUANTITATIVE PCR METHODS

Physiological processes occurring within cells are often profoundly affected by alterations in the level of transcription of key regulatory genes, which themselves alter the expression of other genes. A classical example of this is the *c-myc* oncogene, which is overexpressed in a wide variety of cancers. Quantitation of mRNA transcription may therefore provide clues as to the functional importance of, for instance, developmentally regulated genes. Unfortunately, two main problems arise in this type of analysis. Firstly, the range of transcription levels of a gene may be below that detectable by conventional RNA hybridization techniques (Northern or dot blots). Secondly, biologically relevant data may be obtainable only from living tissues, e.g., mouse embryos or punch biopsies, or specific cell types isolated by fluorescence-activated cell sorting in which case the quantity of material may be minute and the yield of mRNA too low for practical use.

Quantitation of circulating virus particles may be of great value in monitoring the efficacy of antiviral therapies. However, in certain infections, such as HIV or hepatitis C, viremia may be undetectable by conventional molecular or immunological means.

The advent of PCR raised the question whether this technique may be used as a tool for the quantitation of small numbers of nucleic acid sequences. This has led to the development of a number of PCR-based quantitation techniques.

The major area of difficulty associated with PCR as a quantitative method is that no other technique yet matches its sensitivity and can be used to compare results. Therefore, any data obtained by quantitative PCR need to be supported by carefully designed controls and compared with reliable standards. Three main factors affect PCR quantitations.

Polymerase Chain Reaction (PCR): The Technique and Its Applications, written by Rosalind A. Eeles, M.A., M.B., B.S., M.R.C.P., F.R.C.R.; Alasdair C. Stamps, Ph.D.; © 1993 R.G. Landes Company.

TEMPLATE QUALITY

In quantitative PCR, nucleic acid samples are always compared to the relevant standard and, where applicable, to one another. It is therefore important that the PCR template has not suffered any degradation prior to amplification. mRNA quality can usually be checked by Northern blot hybridization using a probe for an abundant message such as β-actin, but with small samples such as fluorescence-sorted cells or DNA from fragments of normal tissue this may not be possible. If samples have been flash-frozen in liquid nitrogen immediately after excision or treated by normal, noncytolytic tissue culture procedures, careful preparation will yield high-quality nucleic acids which, for the purposes of PCR amplification and/or reverse transcription, are undegraded.

THE PLATEAU EFFECT

The exponential amplification potential of PCR accentuates any slight variation in initial reaction parameters. When comparing two templates of different origin, it is therefore essential to show that the efficiency of the amplification of each template is the same. This necessitates multiple sampling at different cycle numbers. As the cycle number increases, amplification efficiency is further affected by substrate saturation and reagent dilution effects. This is the 'plateau effect', so named because the declining copy rate in later cycles produces an asymptotic curve. The point at which the plateau effect occurs will vary with template input and initial amplification efficiency and therefore cannot be predicted for any unknown concentration of template.[1] Thus the range of cycle numbers over which the amplification rate is logarithmic must be determined for each sample. This allows amplification efficiency to be plotted as a function of *log* (product output) against *log* (template input) or cycle number (Fig. 1). During the logarithmic amplification phase this will give a

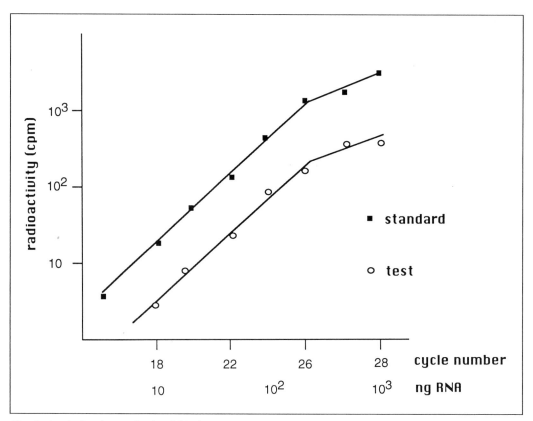

Fig. 1. Analysis of quantitative PCR data.

straight line the gradient of which represents amplification efficiency which may be compared to that of a known standard.

THE TUBE EFFECT

For reasons which are not entirely clear, a single PCR 'master mix', aliquoted into separate tubes and subjected to the same PCR run, will produce a variable final yield of products.[2] Presumably this is due to small variations in a number of parameters, such as sampling errors, temporary local substrate concentrations, tube 'edge effects', temperature across the heating block of the thermal cycler, etc., which combine over many PCR cycles to affect different reactions on a disproportionate scale to their individual contributions. Thus any standard sequence used for comparison in quantitative PCR must be amplified in the same tube as the sequence of interest to ensure, as near as possible, identical reaction conditions. This means that the PCR product derived from such a standard must be distinguishable from that amplified from the test sequence. A number of alternatives have been devised to satisfy this criterion.

ASSESSING PRODUCT YIELD

It is important in a quantitative PCR assay to measure accurately product yield from each PCR run. It is not usually possible to obtain sufficiently accurate data by scanning ethidium bromide fluorescence from a DNA band in an agarose gel. One of two options is normally chosen. The first is to transfer the PCR products from a gel onto nylon membrane by Southern blotting and hybridize the immobilized DNA to a radioactive prove specific to the sequence concerned. The degree of hybridization is then measured by densitometry from an autoradiograph of the blot.[3,4] The other option is to label the PCR products themselves, either by transferring a radiolabelled phosphate group to the 5' end of the oligos using polynucleotide kinase,[5] or by including 2-5 µCi of ^{32}P-labelled dCTP in each PCR tube.[6,7] The addition of this quantity of radiolabel does not significantly alter the concentration of the relevant nucleotide in the reaction.

The use of labelled nucleotide is more likely to produce consistent results as it is commercially available in standardized specific activities, while the labelling efficiency of primers may vary between experiments. PCR products are first visualized by gel electrophoresis and then quantitated by cutting out a slice of the gel containing the band of interest and measuring its radioactivity by scintillation fluorography. Alternatively, a nonradioactive labelling method may be used, such as the incorporation of biotinylated nucleotides or primers.[8] This is an effective and safe means of measuring amplification.

Of the two options, PCR product labelling is the most accurate as transfer and hybridization inconsistencies can be a major source of error in Southern blotting.

Three main quantitative PCR methods have been developed with the aim of reducing or avoiding the amplification artifacts described above. All are aimed at mRNA quantitation but are readily adaptable for DNA samples.

The 'Endogenous Standard' Assay

This method utilizes an endogenous mRNA sequence which is expressed at a relatively constant level in all samples. The level of the 'test' sequence is then compared to this standard.[6,9-13] It is the least laborious of the quantitative PCR methods as the standard sequence is already present in the sample and thus tends to be used when a large number of samples are to be quantitated. It is particularly useful for the study of well-characterized cell lines in which the level of the standard sequence may be confirmed by other means, e.g., Northern blot hybridization. The method is, in fact, the PCR equivalent of Northern blot quantitation in which the expression levels of genes such as β-actin and GAPDH are used to standardize results.

mRNA samples are first converted into single-stranded cDNA by reverse transcription using standard methods.[14] Second strand synthesis is not necessary for PCR as this will be carried out for the oligo-specified sequences in the first polymerization step. The 'window' of logarithmic amplification is then defined for both the test and standard sequences by subjecting a selection of samples to a range of PCR cycles in the presence of oligos for both sequences and a labelled dNTP. To overcome the 'tube effect', both sets of primers are included in each tube, resulting in simultaneous amplification of both sequences. Product yield is assessed for each amplimer and each cycle number chosen and the values obtained corrected for the different sizes of the PCR products. The resulting data is plotted on a log scale against cycle number. The logarithmic phase of amplification is identified by a straight line on this graph (Fig. 1).

Within a range of PCR cycle numbers in which both the test and standard sequences are in the logarithmic phase of amplification, the gradients of the two lines are compared. For an accurate result, these should be a close as possible, as the gradient of this curve gives the amplification efficiency. If both lines have the same gradient, the efficiency of amplification of both sequences is the same, and the relative quantities of input cDNA can be compared. Data for each sample taken from a single point on both lines are used to calculate a comparative index, I, using a normalization formula:

$$I = \frac{\text{test } 1}{\text{standard } 1} \times \frac{\text{test } 2}{\text{standard } 2}$$

where 'test 1' and 'test 2' are PCR product data derived from the sequence of interest, and 'standard 1' and 'standard 2' those of the standard sequence, with 1 and 2 signifying data sets from sample and selected control, respectively.[15]

Although a relatively straightforward way of measuring sequence levels by PCR, this technique suffers from a number of disadvantages. One is that it is only ever a comparative technique: actual sequence numbers cannot be determined, and an essentially arbitrary comparative index is produced. While this may be adequate to describe correlative parameters in cells undergoing change, e.g., correlation of P-glycoprotein

expression with multiple drug resistance,[6] it may be more useful to establish actual basal levels and maximum ranges of mRNA expression for different cell types.

Another potential drawback is that the standard sequence may, in fact, vary in its expression level between two different cell types. The extensive phenotypic difference between, for instance, normal cells and aggressively dividing cancer cells may profoundly alter the level of transcription of even the more 'invariant' genes such as β-actin. Data are therefore normalized to such variations. However, there is no guarantee that all gene expression is perturbed to the same extent or even in the same way by variations in cellular physiology. Thus, expression of the endogenous standard may be downregulated in a certain condition while the sequence under analysis remains constant. In these circumstances normalization of the results will suggest that transcription of the test sequence actually increases relative to the normal level.

Potentially the greatest problem associated with this method is that many endogenous standards, e.g. β-actin, GAPDH, may be expressed at levels far higher than a low-abundance mRNA of interest. If this is the case, dilution of cDNA samples may be required to produce a useful standard curve with the appropriate range of cycle numbers.[11] Unfortunately this means that two separate reactions have to be performed per determination, and amplification efficiencies may vary markedly. With highly purified template and a low cDNA sample requirement which does not affect the PCR buffer to a significant degree, these effects may be small, but such a level of purification is not possible with small samples.

Alternative quantitative PCR methods were developed to counter these drawbacks.

The Synthetic Internal Standard Method

If the quantitative standard for the reaction were identical to the test sequence and of known initial concentration, all the problems of the endogenous standard method could be overcome. Such a standard is obviously of no value in PCR as the origins of coamplified products would be indistinguishable. However, if a synthetic standard is designed to carry very slight sequence variations which nevertheless make it readily distinguishable from the sequence of interest, the two sequences may be amplified in the same PCR tube and assayed separately for comparison. These synthetic internal standards[16-18] are commonly used in quantitative PCR.

To construct an internal standard sequence, a cDNA copy of the gene of interest is altered by-site directed mutagenesis[14,19] to create a unique restriction enzyme recognition site. The position of this is selected so that restriction enzyme digestion of PCR products will result in the release of two fragments which are significantly smaller than the original amplimer (Fig. 2).

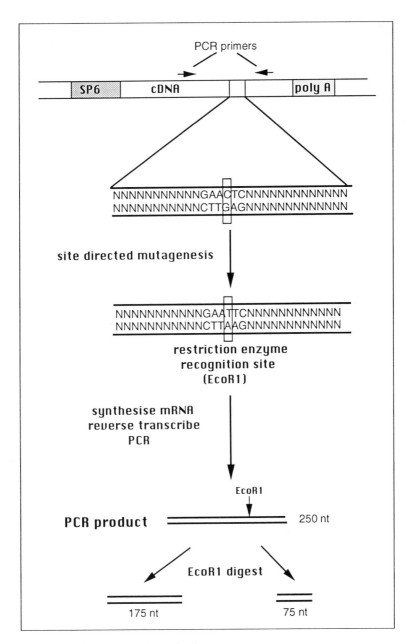

Fig. 2. Synthetic internal standard sequences.

Coamplified PCR products are incubated with the appropriate restriction enzyme and then separated by electrophoresis on an agarose gel. The modified sequence containing the restriction enzyme site will yield two fragments of DNA on this gel, both of which are smaller than the undigested PCR product derived from the normal sequence and may be quantified by the means described above. Thus the internal standard is identical to the test sequence except for perhaps one or two nucleotides. The modified sequence may be cloned into one of the commercially available plasmid vectors which contain a SP6 or T7 polymerase recognition site, directing in vitro transcription of the DNA into RNA (called cRNA). Using this system, large amounts of synthetic cRNA may be produced, enabling accurate weight quantitation by absorbance spectrometry at 260 nm. Using the molecular weight of the cRNA transcript and Avogadro's number, the number of molecules, n, for any given weight of cRNA, R (in grams), can be calculated:

$$n = \frac{R}{M_r} \times 6.26 \times 10^{23}$$

A dilution series of the synthetic cRNA is used to 'spike' RNA samples. Addition of cRNA direct to the samples will also control for efficiency of reverse transcription into cDNA. The amount of cRNA added may be tailored to the range of expression levels of the mRNA under examination to ensure a similar logarithmic amplification phase.

After conversion to cDNA, samples are analyzed using either or both of two assays:

1. product yield versus PCR cycle number at constant RNA input;

2. product yield versus total RNA input at constant cycle number.

The first will delineate the logarithmic amplification phase and may be used as a guide for the number of cycles and range of original RNA input to use for the second assay. A preliminary calibration is required to ascertain the range of cycle numbers over which amplification is logarithmic. Usually, 2-cycle increments from 14 to 30 cycles should give a good range for the average cDNA sample of 30 to 100 ng original RNA. The quantity of cRNA 'spike' may at this stage be found inappropriate and can be adjusted accordingly. Wang et al[20] recommend the addition of a range of cRNA spikes, from ~10^2 to 10^6 molecules per μg of cellular RNA.

When using variable RNA input, each sample should be analyzed as a dilution series, unless the logarithmic amplification range has already been defined. Each sample is diluted by 1:3 for at least six dilutions and amplified for up to 25 cycles. Some dilutions may be too high or too low to fall within the logarithmic amplification phase at the chosen cycle number. Further dilutions or an adjustment of the cRNA spike may be required. Again, a range of spike concentrations may be tried to save time and RNA samples.

Radiolabelled products are analyzed by gel electrophoresis and scintillation counting and the results (the average of at least three determinations) plotted as *log* (radioactivity) against cycle number, and/or *log* (radioactivity) against *log* (input RNA) for both PCR products. A straight line should be obtained, the gradient of which represents the amplification efficiency. Although this may vary between samples, it must be as near as possible identical for the test and standard sequence in each sample. This being the case, the yield of PCR products from the test sequence may be directly compared with that of the standard and converted into molecules of input mRNA. Calculation of RNA copy number from both assays ensures a more accurate result, as does repetition of each experiment in the range of cycle numbers or RNA input giving the best straight line.

This method involves a certain amount of juggling of RNA concentrations and PCR cycle numbers, but results in absolute quantitation of specific mRNA in each sample. In some cases this can be extrapolated to give the number of copies of a particular mRNA in one cell.[20] The strength of the method lies in its greater accuracy as it is based on a standard which is quantifiable by means other than PCR. Detractors of the method claim, however, that the inclusion of an internal standard which is amplified by the same primer set as the sequence of interest sets up competition for resources in the reaction which distorts the true efficiency of amplification. Substrate competition has been exploited in a third quantitative PCR technique.

Competitive PCR

The addition of a set amount of a DNA sequence bearing the same primer annealing sites as the template will divert a proportional amount of primer to its own amplification in a polymerase chain reaction. The new sequence therefore 'competes' with the endogenous sequence for the supply of primers, dNTPs and Taq polymerase at any point in the reaction. Assuming equal amplification efficiency, the degree to which it competes will depend on its initial concentration relative to that of the test sequence. If this is very low, it will scarcely compete at all, resulting in little amplification; if it is high, it may use nearly all the available reagents, allowing almost no amplification of the test sequence. Between these two extremes there will be a point at which the added DNA competes equally with the endogenous sequence, i.e., when the original concentrations of both sequences are equal. This phenomenon has been exploited in the quantitation of DNA sequences by competitive PCR.[3,7,21]

In this technique, PCR is carried out on samples of DNA (or cDNA in the case of mRNA quantitation) in the presence of ^{32}P-labelled nucleotides and varying amounts of a specially mutated competitive standard DNA. Competimer sequences, as these are known, are similar to the synthetic internal standards described under the previous heading in being closely related to the sequence of interest, differing only in either a unique restriction site or a small intron to distinguish their PCR products. cRNA standards are also used to control for efficiency of reverse transcription.[3,21] The range of competimer concentrations must be broad enough to span the point of equivalence with the test sequence. Since this is unknown, Gilliland et al[2] recommend the preparation

of a dilution series from 0.25 to 2.5 x 10^6 attomoles per litre of the competimer, from which predetermined amounts can be added in log increments to aliquots of a master mix of test DNA plus PCR reagents. A finer range of dilutions is used to pinpoint the level of equivalence once it has been approximately defined.

PCR products are separated by gel electrophoresis (preceded by restriction enzyme digestion if necessary) and quantified by scintillation fluorography. Results must be corrected for any molecular weight discrepancies and then expressed as the ratio of competimer to test sequence product yield, averaging at least three samples for each competimer concentration. This ratio is plotted against competimer input. At the point where the ratio of competimer:test PCR products is 1, the corresponding input value will be equal to test DNA (or cDNA) input.

The use of ratios of PCR products for each reaction means that varying amplification efficiencies between tubes can be ignored (although it must be the same for both sequences in any one tube). Because competition between two sequences is being assayed, there is no need to adhere to the logarithmic phase of amplification. However, in the case of competimers containing engineered restriction sites, annealing between single-stranded PCR products representing the two different sequences may occur when oligo concentrations fall too low to allow immediate primer annealing. These heteroduplexes are not recognized by the restriction enzyme, thus increasing the apparent number of PCR products derived from the test sequence. The use of a minimum cycle number yielding quantifiable results prevents this problem.

The competitive PCR method has the advantage that laborious calibrations controlling for amplification efficiencies may be largely omitted, while retaining all the advantages of using a synthetic and precisely quantifiable internal standard. A useful method employing PCR to engineer restriction enzyme site mutants for synthetic competimers has been published by Perrin and Gilliland.[19]

CLINICAL AND RESEARCH APPLICATIONS OF QUANTITATIVE PCR

The major area of usefulness of PCR in the broad area of molecular biology is in the observation of changes in very low levels of nucleic acids which nevertheless have important biological or clinical implications. The literature suggests that this is indeed a powerful and sensitive technique; however, it must be borne in mind that the method is still new and the theory poorly understood, so confirmation by alternative techniques should be sought wherever possible.

HIV: Molecular Virology and Clinical Monitoring

PCR-based detection has immediate relevance to the problem of low levels of viral nucleic acid in HIV and other viral infections such as hepatitis C.[22] Quantitative PCR was immediately seen as a tool which could be used to monitor the HIV status of AIDS patients[23] and to assess the efficacy of antiviral therapies. It also permitted some biologically relevant experiments to be carried out in the dissection of the life cycle of HIV by enabling the study of low-level transcription of viral regulatory genes.[24] The literature concerning AIDS and PCR is overwhelming: this section is not an attempt at full documentation but merely presents a nonexhaustive review of HIV-related applications of quantitative PCR.

MOLECULAR VIROLOGY

Quantitation of nucleic acid levels is particularly relevant to study of the replication mechanism of a latent virus such as HIV which utilizes low-level regulatory proteins to maintain its genome in an apparently inactive 'stand-by' state for long periods of time. Conventional molecular techniques applied to virology require the high expression of viral genes in cell culture systems, but this is liable to produce misleading results. Furthermore, the tight arrangement of viral genes leads to complex RNA splicing to enable independent expression of viral proteins; these transcripts may be difficult to distinguish by conventional Northern blotting, while PCR is able both to distinguish and quantify differentially spliced messages. Recognition of this potential led Arrigo and colleagues[24,25] to use full wild-type and deleted genomic clones of HIV to demonstrate the role of the HIV *Rev* gene in viral replication and latency. This group also exploited PCR quantitation and specificity to show that reverse transcription of the HIV genome occurs at comparable levels in both quiescent and activated T lymphocytes but that viral replication is limited in quiescent cells by failure to complete full-length transcription.[26] Further evidence suggested a role for this incomplete DNA in the production of infective virus in activated T cells,[27] a finding with potential relevance for drug targeting.

More clinically oriented research aimed to correlate the level of viral mRNA expression with stages of HIV infection and AIDS.[28] Other researchers have investigated cellular transcriptional responses in HIV infection, such as the coordinated cytokine response of HIV-infected promonocyte cell lines to subsequent antigenic challenge.[29]

Schnittman and colleagues[30] reported a 4- to 10-fold higher rate of HIV infection of CD4+ memory T cells of naive CD4+ cells in vitro using a quantitative PCR technique to detect viral sequences post-infection. Studies on CD4+ T lymphocytes from patients at different stages of HIV infection suggested that higher copy numbers of the HIV DNA provirus were present in patients with AIDS than in asymptomatic carriers.[31]

A quantitative modification of PCR was used to demonstrate the presence, at high levels, of the three main forms of HIV DNA in brain tissue of AIDS dementia sufferers.[32]

Rich and colleagues[33] showed that differentiated mononuclear phagocytes isolated from lung alveoli were considerably more susceptible to HIV infection than fresh blood monocytes. This particularly thorough investigation examined all HIV transcripts as well as intermediate and double-stranded viral DNA by quantitative PCR, demonstrating that differences in levels of productive infection of macrophages at different stages of cellular maturation may be due more to variation in the ability of the virus to enter cells than failure to replicate the viral genome.

CLINICAL STUDIES

The use of quantitative PCR in clinical applications is still at an early stage due to the previously mentioned problems of standardization and control. Nevertheless, a large number of publications may be found in which quantitative PCR has been used to monitor peripheral blood HIV load. A selection of these show that it can be a useful method in prognosis and therapy.

Early reports intimated some concern about the feasibility of quantitative PCR as a diagnostic technique which could not be confirmed by a second approach.[34-36] Improvements in methodology, particularly with regard to controlling for amplification efficiencies using another endogenous DNA standard,[12,37] synthetic internal standards[17,38] or competitive PCR,[39,40] increased confidence in results. Most quantitative studies on HIV relate to the monitoring of antiviral therapies such as AZT,[31,41,42] dideoxyinosine[34] and dideoxycytosine.[43] Studies on the activity of antiviral drugs have also been carried out in vitro on primary HIV1-infected T cells.[44]

Attempts have been made to assess viral load in patients as an indicator of disease status. For example, in follow-up studies on two cohorts of 12 and 15 HIV positive individuals, a correlation was established between disease progression and an increased proportion of HIV infected, CD4+ peripheral blood T lymphocytes.[45] An inverse relationship between proviral burden and CD4+ T cell count was also documented by Oka and colleagues.[36] A large study of HIV1 DNA copy number in peripheral blood mononucleocytes was carried out on patients at various clinical stages.[46] Daar et al[47] observed rapid declines from high levels of viremia (plasma and peripheral blood mononucleocytes) during acute phase infection in four patients, suggesting an immune response during this period. A peak of viral DNA load at the time of p24 seroconversion was documented by Jurriaans and colleagues.[39]

Immunology

The extent of knowledge of cell population dynamics and differentiative responses of the hemopoietic system present a number of questions to which quantitative PCR may be applied. The method has been directed particularly at two areas: cytokine response and the localization and quantitation of T cell subpopulations.

Some priority has been given to establishing the reliability and, in some cases, verification of the method. An endogenous standard mRNA, dihydrofolate reductase, was used to control for a study of TNF-α expression patterns in draining lymph node cells as a response to antigenic stimulus.[48] Kanangat and coworkers[49] utilized the system of Wang et al,[20] in which a synthetic standard was produced by joining sets of cytokine-specific primer pairs together in an in vitro transcription vector. The primer sequences were arranged in such a way that a primer pair specific to any particular gene could amplify a sequence of predetermined length from the standard which would give a PCR product of a size distinguishable from that of the endogenous mRNA. Reproducible measurements of mRNA levels for IL-2, IL-6 and TNF-α mRNA levels were thus obtained from Jurkat, THP-1 and HL60 hemopoietic cell lines. Platzer and colleagues[50] used this type of construct in a competitive PCR system to assess the effect of IL-4 production on the expression of the other interleukins, interferon gamma and the IL-4 receptor in an inducible IL-4-transgenic mouse. A study of macrophage response to IL-4-producing cells in the nude mouse revealed an up-regulation of mRNAs encoding IL-4 receptor, IL-5, TNF and interferon γ.[51] The competitive PCR assay of Becker-Andre and Hahlbrock[3] was used to assay phorbol ester-induced cytokine response in a leukemic cell line. Stimulation of plasma cell proliferation by adherent bone marrow cells expressing high levels of IL-6 in multiple myeloma patients could be reduced, and IL-6 production blocked, by IL-4 in vitro. The expression of high levels of IL-6 in adherent bone marrow cells of multiple myeloma patients could be blocked by IL-4 in vitro, with a concomitant reduction in plasma cell proliferation, suggesting a therapeutic role for IL-4.[52]

In Crohn's disease, higher levels of IL-2 mRNA were detected in areas of inflammation as compared to normal tissue, peripheral blood lymphocytes or samples from patients with ulcerative colitis,[53] suggesting a mechanism of T cell activation in the pathogenesis of this condition. Quantitative PCR has also been used to address the role of IL-7 in mouse embryonic thymus development, demonstrating a peak of expression in 15-day embryos which was confirmed by in situ immunofluorescence studies.[54] In cases of idiopathic pulmonary fibrosis, elevated expression of IL-8 by lung alveolar macrophages and its correlation with increased numbers of neutrophils supported evidence for the involvement of these cells in disease progression and their activation by macrophages.[55] The production of mRNA encoding the cytokine synthesis inhibitory factor, IL-10, by keratinocytes, normalized to β-actin levels, was shown to increase within 4 hours of the application of certain chemicals, including contact allergens, to mouse skin.[56]

The clonality of a population of T lymphocytes is indicated by the range and levels of expression of the T cell receptor (TCR) V-β genes. Quantitative PCR may therefore provide information on the clonality of small T cell populations in, for instance, pancreas or synovial fluid samples. In a mouse model of heritable diabetes, the predominant expression of the V-β11 transcript in pancreatic T lymphocytes of younger mice suggested a role for this subpopulation in the onset of insulitis leading to heritable diabetes.[57] T cells contained in synovial fluid taken from 18 patients with juvenile rheumatoid arthritis exhibited a restricted TCR V-β gene repertoire.[58]

Monitoring of the early posttransplantation phase of bone marrow transplants could be carried out in a reasonably rapid quantitative PCR assay, assuming a high degree of experimental reproducibility. One such study, by Roth and coworkers[59] showed that residual host cells persist for 1 to 2 weeks posttransplantation, with subsequent rapid disappearance.

Studies on human intestinal intraepithelial T cells demonstrated oligoclonality which was nevertheless restricted in TCR repertoire compared to peripheral blood lymphocytes.[60,61] Another report documents the evolution of memory B cells during childhood, using flow cytometric separation of the cells followed by PCR quantitation of defined V-H gene rearrangement.[62]

Cancer

The use of quantitative PCR in clinical cancer may provide a range of new prognostic markers and indicators of disease stage by targeting low-level expression of key regulatory genes involved in cancer cell phenotypes. To some extent the method is, in this context, still an answer waiting for a question, but some areas of application have already been found.

Management of certain leukemias could be extended using quantitative PCR, as in a study on patients with chronic myelogenous leukemia, in which quantitative detection of the *bcr/abl* rearrangement was used to monitor residual disease after bone marrow transplantation and interferon alpha treatment.[63]

The point at which anticancer drugs such as antifolates and fluorinated pyrimidines act in biosynthetic pathways could be examined by studying transcriptional levels of the genes involved, as in a study on dihydrofolate reductase and thymidylate synthase levels in tumors of patients treated with such drugs.[64] Expression of the estrogen-induced pS2 gene in a human breast cancer cell line was shown to be repressed by retinoic acid.[65]

The recommencement of fetal protein expression patterns is a feature of many dedifferentiated tumor cells. For instance, the enzyme sucrase-isomaltase was found to be over-expressed in 80% of 30 colonic adenocarcinomas when compared to levels in the surrounding normal mucosa, in which it is either not expressed or transcribed at a very low level.[66] Transcription of the gene encoding human transcription factor IID was elevated relative to a range of normal tissue types, including lung, in almost all lung carcinomas and cell lines examined in a study by Wada et al.[67] In contrast, a gene expressed from a region of human chromosome 3p1 was transcribed at only 3% of normal levels in small cell lung carcinoma.[68] Defining specific levels of gene expression in tumor cells by PCR may thus assist in identification of tumor types and stages.

Expression of the multiple drug resistance genes, specifically the P-glycoprotein-encoding MDR1 gene, has therapeutic significance in human tumors. An extensive analysis of normal tissues, tumors and tumor cell lines with respect to MDR1 transcription level was carried out by Noonan and coworkers[6] using the endogenous β-microglobulin mRNA as a standard for membrane protein expression. MDR1 transcript levels correlated with P-glycoprotein expression and drug resistance, while MDR1 negativity was most common among tumors which were responsive to chemotherapy.

Other Applications

Hormonal responses

Cellular responses to hormones are manifested by alterations in transcriptional activity affecting the expression of genes involved in regulatory or biosynthetic pathways. Quantitative PCR can be used to assess these transcriptional effects which often occur within ranges which are undetectable by conventional means. The rapid response of the cell cycle regulatory genes *c-fos* and *c-myc* to gonadotropin stimulation in rat ovarian granulosa cells was evaluated by PCR quantitation of mRNA and confirmed by immunocytochemical staining of the transcribed proteins, suggesting a molecular pathway leading to proliferation and differentiation of these cells.[69] The levels of FSH, LH and progesterone receptors in ovarian granulosa and theca were also assessed by quantitative PCR in a study of the response of these genes PMSG and hCG.[70,71]

Developmental biology

Fluctuations in the levels of certain genes signal, and may affect, stages in development. The most direct effects may be seen in early embryos with the result that quantitative PCR may be the only method applicable to the tiny amounts of mRNA present in the developing structures. Studies of this kind have been carried out on nerve growth factor-related genes;[72] genes involved in spermatogenesis, i.e., HPRT, phosphoglycerate kinases I and II, the Y-chromosome linked Zfy gene, adenine phosphoribosyl transferase and cytosine-5-methyl transferase;[73] the thyroid hormone receptor gene, c-erbA-alpha;[74] the muscle determining factor, MyoD[75] and the lens membrane protein, calpactin I.[76]

Hereditary diseases

Quantitative PCR has been used in diagnosing sex chromosome aneuploidy[77] and in identifying carriers of Duchenne and Becker muscular dystrophy by simultaneous amplification of multiple exons of the dystrophin gene.[78,79]

Hypomethylation and gene expression

Hypomethylation of chromosomal gene regulatory sequences may lead to overexpression of genes in the near vicinity. It may be assayed by initially digesting total genomic DNA with a frequently-cutting, methylation-sensitive restriction endonuclease. DNA which remains uncut is then quantified by PCR using oligos specific to the site of interest. Correlations have been observed between hypomethylation of potential regulatory DNA sequences and increased gene expression from the same region.[80,81]

USING PCR TO ANALYZE THE FLANKING REGIONS OF KNOWN DNA SEQUENCES: INVERSE AND VECTORETTE PCR

DNA amplification by PCR depends on the presence of two suitable primers in the reaction mix, the design of which requires precise knowledge, at the nucleotide sequence level, of primer annealing sites in the region to be amplified. At first sight, this statement appears to imply that only known sequences may be amplified by PCR. However, this is not necessarily the case, as a number of ingenious methods have been devised to enable limited sequencing of DNA flanking a known sequence. These methods are capable of yielding important information, such as details of a chromosomal breakpoint in a leukemia or other cancer, the site of chromosomal insertion of a cancer-associated virus, or the 5' end of a cDNA, enabling its expression in vitro and the study of its encoded protein.

INVERSE PCR

The first method designed to amplify unknown DNA sequences by PCR was presented by Ochman and coworkers.[82] In the general application of this technique (Fig. 3), DNA containing the sequence of interest is digested, using a suitable restriction enzyme, to produce a 2000 to 3000 nucleotide restriction fragment containing the known 'target' sequence flanked by two regions of unknown sequence. The size of the fragment is important as it must represent a reasonable length of DNA for PCR amplification. The source of the DNA may be anything from chromatin to cDNA to the purified insert of a lambda bacteriophage vector. The specificity of PCR ensures that only the targeted sequences will be amplified. After digestion the DNA is diluted to a low concentration. This favors the formation of monomeric, circular molecules when the DNA is ligated, as it is in the next step. Defining this concentration in a population of mixed fragments is difficult, and a calibration experiment may need to be carried out to find a concentration which gives good results. The ligated DNA circles are then amplified by PCR. Two oligos are designed which anneal to opposing strands but which direct Taq polymerase-catalysed DNA synthesis away from one another, in contrast to the usual PCR amplification. For this reason, the method was dubbed 'inverse' PCR. New DNA strands primed by either oligo extend round the circularized DNA, eventually incorporating a complementary copy of the other primer. A linear strand of DNA is thus produced, as the Taq enzyme is unable to displace the annealed primer as it attempts to make a second round of the circular template. This new DNA contains opposing primer sites and is now a bona fide PCR template which is amplified in the usual way (Chapter 2). The first PCR cycle thus effectively cuts the circularized molecule between the two primer sites, producing an amplifiable DNA fragment consisting of the two 'unknown' flanking regions joined end-to-end. Since Taq polymerase works slightly more efficiently with linear than circular DNA, the circular fragments may be linearized before amplification by digestion with a restriction enzyme which cuts only at a position between the two primers in the known target DNA sequence.[83] After PCR, sequencing of the amplified DNA is carried out using one or other of the original oligos (or oligos annealing slightly within the known region of the amplified sequence) as the sequencing primer. A 'chimera' of the two flanking sequences results, but the ends of these sequences coincide with the original restriction enzyme recognition site, which was reconstituted when the digested template DNA was circularized by ligation. The flanking sequences may therefore be distinguished from one another by noting on which side of the restriction site they lie relative to the sequencing primer used.

Inverse PCR is a rapid and reliable method for obtaining sequences flanking a known region of DNA. Novel sequences of 1000 to 2000 nucleotides may be obtained using consecutive sequencing primers, and having sequenced these regions more may be analyzed by the same method using suitable restriction enzyme digests and new primers. However, 'chromosome crawling'[83] in this way is a very expensive way of obtaining sequence compared to conventional subcloning techniques, as a large number of sequencing and PCR primers are required. Inverse PCR is therefore best applied in situations where a short sequence will yield important information, such as in the analysis of viral integration junctions,[84,85] or chromosomal translocation breakpoints,[86-89] or sequencing the 5' ends of cDNAs.[90,91] Its major limitation is that a restriction enzyme recognition site must be found within the flanking regions which does not occur within the target sequence and which will produce, after restriction enzyme digestion, a fragment containing significant amounts of both flanking regions, but not so much that PCR amplification of the circularized molecules is inhibited or prevented. This presupposes a detailed knowledge of the available restriction sites within the flanking region. In the case of cDNA analysis,[90,91] in which no restriction digestion is necessary as the blunt-ended linear molecule is simply self-ligated,

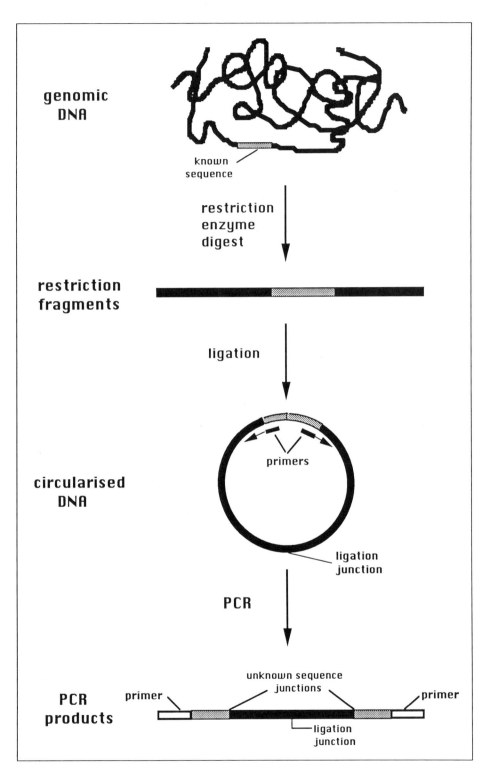

Fig. 3. Inverse PCR.

and the sequence of one end is already known, this poses no problem. Similarly, known, common viral integration sites[84] and chromosomal translocation breakpoints[86,89] in which stable landmarks in the restriction map around the target sequence are known, may be routinely studied by inverse PCR. Other studies, however, charting unknown regions of DNA, have to proceed on the assumption that a suitable restriction site will be found by trial and error. Nevertheless, the method has found wide application in this respect as well.

Applications of Inverse PCR

The integration of viral DNA sequences into the chromosomal material of the host cell occasionally results in the disruption of the unique sequences which make up approximately 10% of the mammalian genome and contain the genetic information. This may cause an alteration of the expression of a cellular gene. In some cases this causes oncogenic conversion, resulting in tumorigenesis. The study of viral integration in tumor cells may therefore reveal one of the causes of tumor formation. Inverse PCR followed by direct sequencing is a rapid method of analysing an integration site, provided only one such site exists in the tumor DNA. If there are more than one, the products of inverse PCR must be individually cloned and identified, as direct sequencing would result in the superimposition of two or more sequences in the final sequencing gel.

Integration of the Friend murine leukemia virus at a common site in erythroleukemic spleen cells of experimentally infected mice correlated with the finding that 75% of murine erythroleukemic cell lines contained chromosomal rearrangements at this site.[84] Similarly, a natural strain of the feline leukemia virus which incorporates and drives expression of a cell-derived *myc* oncogene was found to integrate at a common locus in lymphosarcomas which developed in experimentally infected animals. This integration was found, through inverse PCR techniques, to be altered in many lymphosarcomas of nonviral origin,[85] suggesting that the disruption created an oncogene which could cooperate with the virally over-expressed *c-myc*

to cause tumor formation. A study of integrational mechanisms underlying the repeated duplication of cellular sequences encoding a transfer RNA in the newt was carried out by generating sequence from flanking regions of known DNA targets through inverse PCR.[92]

Another important area of application of inverse PCR is in the study of chromosomal translocation breakpoints. These often occur in predictable regions of specific chromosomes in certain types of cancer. In order to analyse the genes involved, it is necessary not only to clone the DNA containing the breakpoint but to examine the normal, unrearranged chromosomes which originally took part in the translocation event. Inverse PCR may be used to obtain sequence information and clone informative DNA fragments at all stages, since the analysis of breakpoints requires extensive mapping at the level of restriction endonuclease recognition sites and the identification of known sequences close to either side of the breakpoint.[88] At a more preliminary level also, inverse PCR may be used to characterize the ends of long cosmid or yeast artificial chromosome (YAC) inserts, since these will, of course, flank the known vector cloning site sequences.

Inherited diseases such as Duchenne and Becker muscular dystrophy are often caused by germline chromosomal translocations. In one report, the positions of such translocations within the affected dystrophin gene were identified by inverse PCR cloning and the postmeiotic origin of apparently paternally inherited muscular dystrophies demonstrated.[89] The universal genetic basis of certain types of thalassemia in widely disparate populations was shown to be a deletion breakpoint occurring within a very small sequence in the beta-globin gene cluster.[86] Partial deletion of the antithrombin III gene, the basis of a heritable deficiency, was also analyzed by sequencing via inverse PCR in an extended family group.[87] Inverse PCR was used also to investigate the repertoire of T cell receptor expression in mononuclear cells from the synovial fluid of patients with rheumatoid arthritis, using primers annealing to the constant regions flanking the variable receptor sequences.[93]

In cDNA cloning and analysis, it is often difficult to obtain sequences at the very 5' end of the gene. A number of researchers have circumvented this problem by the use of inverse PCR. Double-stranded cDNA may be self-ligated in monomeric circular form for amplification by PCR primers annealing to the 3' and 5' ends of the known sequence. Since the 3' end of the full-length cDNA is signalled by its poly(A) tail, the junction between it and the true 5' end of the molecule is readily established and the unknown 5' end sequences obtained.[90] This adaptation of inverse PCR is, however, sometimes prone to the same sort of problems as those encountered in cDNA cloning in the first place.[91] For instance, difficulty in obtaining 5' sequence may be due to mRNA secondary structure in this region preventing efficient reverse transcription into cDNA. This results in the synthesis of a number of shortened cDNAs which are preferentially amplified by PCR. On direct sequencing of such PCR products, the signal produced by 5' end sequences will be very faint, or if the DNA fragments are cloned, a very small proportion of the clones will represent full-length cDNA leading to a great deal of extra work in eliminating the truncated sequences. Despite these potential drawbacks, inverse PCR has been successfully applied in a number of fields to the cloning of full-length cDNAs and the generation of sequence from the 5' ends of genes.[90,91,94-97]

Vectorette PCR

The limitations imposed upon inverse PCR by the requirement for conveniently placed restriction enzyme recognition sites has led to the development of other methods for sequencing 'unknown' DNA which are not restriction enzyme-dependent. The main obstacle to be overcome is the fact that, in moving away from the inverse PCR technique, only one primer annealing to the known target DNA may be used, and a second primer annealing site must be placed at a distal position to allow amplification of the unknown intervening sequences. Short,

double-stranded DNA 'linkers' may be ligated to the ends of restriction enzyme-digested DNA, allowing amplification between primers annealing to the linker and target sequences, but this would also permit amplification between linkers at each end of any DNA fragment, particularly the shorter ones, swamping the reaction with mixed PCR products. Therefore, a primer sequence must be placed on the ends of DNA molecules in such a way that only those fragments containing the target sequence are amplifiable. Ligation of restriction enzyme-digested DNA into a plasmid vector, if carried out carefully, effectively places a different primer annealing site on either end of each fragment in the form of the vector cloning site. Amplification with oligos specific to the target sequence and one side of the vector cloning site ensures that only those molecules containing the target sequence ligated in the correct orientation may be amplified logarithmically. In practice, however, asymmetric DNA synthesis from every ligated plasmid competes strongly for primers and dNTPs and produces a significant background.[98]

These problems have been overcome by Riley and coworkers[99] using a partially double-stranded linker which is termed a 'vectorette'. The vectorette (Fig. 4) consists of two oligos of around 60 nucleotides in length which contain complementary sequences to one another at either end but not in the middle region of about 30 nucleotides. The two oligos thus anneal at either end but 'loop out' in the central, noncomplementary region. This annealed structure forms the vectorette. One of the oligos also has an extra four nucleotides at one end which, in the annealed vectorette, produces a 'sticky end', compatible with those left by digestion of DNA with a frequently-cutting restriction enzyme such as Hinf I or Rsa I.[99] Finally, a PCR oligo is synthesized to contain the same sequence as the noncomplementary region of the shorter vectorette oligo.

DNA containing the target sequence and its unknown flanking regions is digested with the restriction enzyme specified by the sticky end of the vectorette and ligated to an excess of vectorette. The sticky end ensures that the vectorette ligates to the digested DNA fragments in the correct orientation. This 'vectorette library' is then subjected to PCR amplification using a primer which anneals to the target sequence and the oligo which matches the vectorette noncomplementary region. During the first cycle of the PCR, no primer annealing to any vectorette sequence will occur, as no complementary sequence exists for the vectorette primer. Amplification therefore depends entirely on

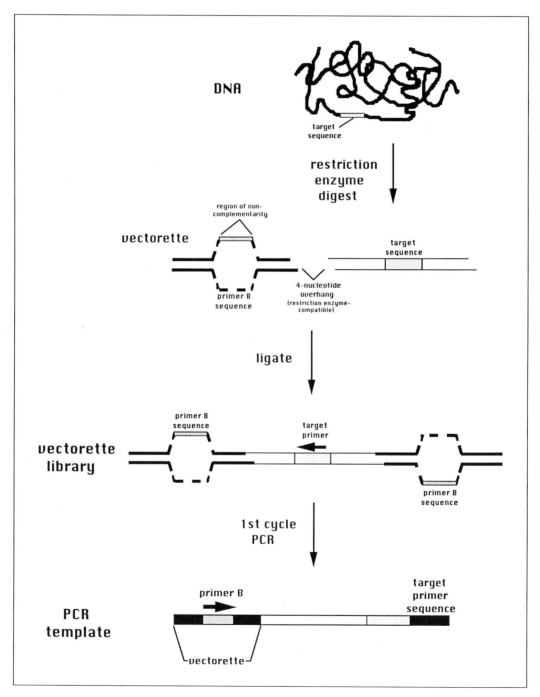

Fig. 4. Vectorette PCR

synthesis of new DNA strands from the target primer which extend past the ligated vectorette primer sequence, to produce a complementary annealing site for this oligo. These molecules, synthesised in the first cycle of the reaction, provide a specific PCR template for amplification. The resulting PCR products may be sequenced using the template oligo as a sequencing primer to obtain flanking sequences.

Vectorette PCR is a potentially very powerful method for obtaining unknown DNA sequence flanking known targets. It is not dependent on any prior knowledge of restriction sites in the unknown flanking region. In fact, any restriction site compatible with the vectorette may be used, and the vectorette may be made compatible with any chosen restriction enzyme.

It is therefore somewhat surprising that very few groups have taken advantage of the technique. It was originally developed as a means of sequencing the ends of inserts of YACs,[99,100] enabling the subsequent ordering and classification of overlapping cosmid clones containing the YAC insert sequences. In this application of vectorette PCR, the sequence of the vector cloning site was used to produce a target oligo. This approach was also used to determine the sequence and thereby map the exon structure of the 3' end of the dystrophin gene, using both the YAC vector cloning site sequences and known 3' exon sequences of the dystrophin cDNA as target sites in a 'genome walking' exercise.[101] Although sequencing of the ends of YAC inserts can be equally carried out by inverse PCR,[102] the vectorette library approach offers the advantage that circularization conditions do not need to be established for the ligation step.

The application of vectorette PCR to the analysis of chromosomal breakpoints was demonstrated by Mills et al[103] in an analysis of the *bcr-abl* chimeric gene formed by translocation in chronic myelogenous and acute leukemias. An adaptation of the method, in which straightforward double-stranded oligonucleotide linkers were ligated to restriction enzyme-digested genomic DNA, was used by Espelund and Jakobsen[104] to clone and sequence novel transcriptional promoter elements flanking known gene sequences. Linear amplification was first carried out using a biotinylated primer which annealed to the target sequence. The resulting single-stranded molecules were then captured using streptavidin-coated magnetic bead technology, removing the original genomic DNA and excess linkers. The captured DNA was then amplified using the target primer and an oligo annealing to the ligated linker sequence and cloned and/or sequenced directly (see below: "PCR sequencing methods").

ISOLATING GENES BY PCR AMPLIFICATION WITH DEGENERATE PRIMERS

The molecular approach to studying biochemical activities and functions is to clone the gene encoding the relevant protein and express it in an appropriate assay system. Conventional gene cloning strategies involve the isolation of specific sequences from a cDNA library, either:

(a) by using a DNA probe derived from a larger functional or genetically informative genomic sequence, e.g., a YAC or cosmid; or

(b) via a peptide sequence obtained from the purified protein, using oligodeoxynucleotide probes designed to represent all DNA sequences which could potentially encode that peptide.

The latter approach is now increasingly taken as new peptides are identified through functional interaction with known proteins, and as peptide sequencing has become increasingly refined, requiring only very small quantities of protein. The DNA sequence coding for a particular peptide may be predicted using the universal genetic code; however, the code is redundant for all amino acids except methionine and tryptophan, and up to six different codons may specify just one amino acid, (e.g., arginine; see Figure 5). Oligos designed to probe for coding sequences are correspondingly highly degenerate. For example, a pentapeptide sequence, Cys-Tyr-Ser-Arg-Pro, taken from the glycoprotein B gene of the herpesviruses,[105] could potentially be encoded by over 300 different 15-nucleotide oligos (Fig. 5). Although this number can be reduced by observing codon usage bias in the organism concerned, degeneracy still remains high enough to affect the specificity of the oligo as a hybridization probe. As a result of this, either the degenerate oligo mix may contain too few specific sequences to give a hybridization signal on the true gene sequence, or hybridization stringency may be so far reduced that a large number of nonspecific cDNAs are identified which have to be sequenced and rejected before the targeted gene is isolated.

Lee and colleagues[106] used two degenerate oligos as PCR primers to generate a DNA fragment of reasonable length for use as a specific hybridization probe for porcine urate oxidase. The oligos were based on the short stretches of the first 32 amino acids of peptide sequence obtained from 100 picomoles of purified porcine urate oxidase, and included every codon degeneracy. Opposing five-amino-acid sequences were chosen from either end of this sequence, and the first two nucleotides encoding the sixth amino acid were included in the N-terminal primer. This gave added specificity to this primer, as the first two nucleotides of most codons are nonredundant. The primers also contained nonspecific nucleotide sequences at their 5' ends which, when converted to double-stranded PCR products, produced restriction enzyme recognition sites. This enabled PCR products to be cloned into plasmid vectors after appropriate restriction enzyme digestion (Fig. 6). A third degenerate oligo, with minimal homology to the PCR primers, was designed to hybridize to a short sequence within the expected specific PCR product.

peptide	Cys	Tyr	Ser	Arg	Pro
oligos	TGT	TAT	TCT	CGT	CCT
	C	C	**AGC**	**A** C	C
			A	**A**	A
			G	G	G

(exclusive nucleotide combinations in bold)

Codon degeneracy

amino acid	no. of codons
M,W	1
C,D,E,F,H, K,N,Q,Y	2
I	3
A,G,P,T,V	4
L,R,S	6

Fig. 5. Degenerate oligos.

This was used as a probe to confirm that the correct sequence had been amplified, and to pick plasmids containing this sequence after cloning of the PCR products. Plasmids containing urate oxidase sequences (further confirmed by sequencing) were themselves used as probes to isolate a full-length cDNA sequence from a porcine liver cDNA library.

This method demonstrated that a pair of degenerate oligos was capable of amplifying cDNA with a remarkable degree of specificity, the major PCR product being the expected urate oxidase sequence, despite the presence of hundreds of different primer sequences. Because successful PCR amplification requires two primers, the degeneracy of

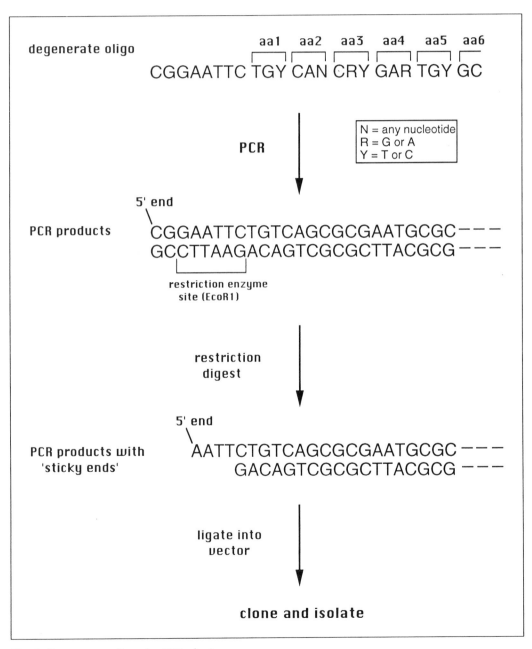

Fig. 6. Degenerate oligos for PCR cloning

the oligos is compensated for. Thus, although many otherwise nonspecific stretches of DNA sequence may contain annealing sites for the various different oligos present in the reaction mix, the probability of two opposing primers being close enough to allow efficient amplification is very low. In contrast, full oligo degeneracy ensures that primers are available for the specific priming sites in the cDNA or gene of interest. This results in a considerably higher overall specificity than can be obtained by single-oligo probing of cDNA libraries. The use of a third degenerate oligo as an internal probe for specific amplification products further increases the likelihood of identifying the correct gene before laborious sequencing procedures need to be carried out. Again, this is an improvement on conventional cDNA library probing with a degenerate oligo as the nonspecific background is reduced by many orders of magnitude.

For most proteins, limited amino acid sequencing should produce a unique sequence, allowing specific cDNA amplification even with degenerate oligonucleotides. In some cases, however, this is not sufficient to obtain a single, specific PCR product. If the equivalent DNA sequence distance between the two chosen peptide sequences is known, the specific PCR product may be selected on size and by hybridization to a third oligo,[106] but such intervening peptide sequences are not always available. The prospect of cloning and sequencing a large number of different DNA fragments may be daunting, and in such cases more refined approaches have been developed. Strub and Walter,[107] in cloning the 14 kilodalton subunit of the signal recognition particle (SRP14), used one degenerate 'reference' primer with a series of opposing primers derived from a long, distal peptide sequence. PCR products were derived from cDNA using this set of primers, and then compared to amplified DNA from another, overlapping set consisting of a different reference primer with the same opposing primers. Comparison was carried out by Southern blot hybridization: one set of PCR products was separated by gel electrophoresis and immobilized on nylon membrane, then probed using another, radioactively labelled set. Using this approach, those sequences which were amplified by both sets of primers could be identified and isolated. Out of over 40 major PCR products, two were picked out by this cross-hybridization method. These DNA fragments were then cloned into plasmids, further purified by hybridization to labelled PCR products, and sequenced. Limited sequencing only was required to identify clones containing SRP14 gene sequences, which could then be used to pull out a full-length cDNA. A similar approach was also used successfully by Gonzalez and coworkers[108] to clone a cDNA coding for a protein binding to the genomic cAMP response element.

Amplification with degenerate oligos introduces a set of new PCR parameters which must be set or calibrated for efficient and specific amplification. Lee et al[106] noted, on sequencing a number of specific cloned PCR products, that all contained a few mismatched nucleotides in the original primer sequences. These mismatches were generally confined to the 5' ends of the primers and therefore did not interfere with specific annealing at the crucial 3' end during amplification. The annealing temperature used in these experiments was low, at 28°C, thus allowing a higher degree of mismatching. This may be reduced by raising the stringency of the annealing step[105] after the first few cycles by setting the annealing temperature at an optimum based on the T_m of the degenerate oligos (see Chapter 3). Without the aid of a computer program, this may involve an extremely long calculation, so a few calibration experiments involving different annealing temperatures may be a more straightforward option. If amplification proves difficult, the ramp time between annealing and polymerization may need to be increased to promote annealing efficiency, although it may also encourage nonspecific annealing. Careful selection of amino acid sequences, avoiding repeat sequences and choosing codons with low degeneracy where possible, tends to increase specificity.

Another factor to be borne in mind when using degenerate oligos for PCR is that each individual sequence is present at only a tiny fraction of the total oligo concentration. Since degeneracy can push the total number of different sequences for one coding oligo into five figures even for a 7 amino acid peptide,[109] this fraction may be very small indeed, in this case representing <1 femtomole. Although this may not limit the initial stages of amplification, the 'plateau effect'[110] will occur much earlier than usual and well before an observable quantity of specific PCR products are present in the reaction. At this stage and with low annealing temperatures, oligos with single nucleotide mismatches, particularly at positions away from the 3' end, are able to compete for annealing to the specific template. As a result, many, if not all PCR products isolated and cloned into plasmids will contain nucleotide mismatches at either end when finally compared to the true gene sequence.[106] While these may have no implications for the subsequent use of such sequences as probes, the efficiency of the original amplification may be compromised, leading to difficulty in obtaining sufficient amplified DNA for cloning. Adding large amounts of degenerate oligos, i.e., up to 10 μmol[111] in place of the usual 20-100 pmol, together with calibration (or calculation) of the annealing temperature contributes to amplification efficiency by promoting specific primer annealing.

Other researchers have proposed the use of deoxyinosine to substitute for nucleotides at ambiguous positions in degenerate codons. Deoxyinosine binds in complementary fashion and with equal affinity to all four naturally occurring nucleotides in an opposing DNA strand. This approach has been demonstrated to be successful in amplifying a gene of known sequence[112] and was used by Patil and Dekker[113] to amplify part of the 2-keto-4-hydroxyglutarate aldolase gene of *E. coli* in a single PCR product. The advantage of this modification is that only two oligo sequences are present which should anneal with only moderately reduced affinity to the specific primer sites. However, too high a degree of degeneracy will increase the probability of nonspecific amplification which, in the same way as with normal PCR, will occur with high efficiency once the template

has been copied. Therefore strict preference must be given to amino acids with limited codon usage, particularly in the 3' regions of the oligos.

In cases where a single major PCR product is obtained using degenerate oligos, direct hybridization of radiolabelled amplified DNA to cDNA libraries may yield specific sequences. This 'short cut' approach has been used successfully to isolate a cDNA encoding the tumor necrosis factor receptor.[114]

CLONING GENES ENCODING RELATED PEPTIDE SEQUENCES

Many genes found in complex organisms are part of a 'gene family' encoding proteins which have related functions. These functions are mediated by active sites within catalytic domains of the translated protein: short sequences of amino acids which, in the active protein, form a 'pocket' or similar structure in which specific biochemical reactions are catalysed. Often the same reaction is utilized by the living cell in a number of different contexts. These reactions, such as targeted phosphorylation and dephosphorylation, proteolytic activity, oxidation and reduction, are fundamental to inter- and intracellular communication, cell division and energy transfer. Perhaps it is not surprising, therefore, that proteins which catalyze the same reaction should have tightly conserved amino acid sequences at their active sites, while the conditions under which they become active are interpreted by other, nonconserved regulatory domains. Conservation of active site sequences provides a handle which biologists can use to dissect cellular mechanisms and biochemical pathways, leading in the clinical context to a greater understanding of disease states and opening up the potential for targeted drug design.

The most broadly studied gene families are the protein kinases which catalyse the transfer of phosphate groups onto specific amino acids, either on another protein or within their own peptide sequence. Phosphorylation may activate or repress certain protein functions. Kinases are thus involved in cellular signalling pathways, e.g., the cellular insulin response, and some are proto-oncogenes.

Potential oncogene status has focused attention on the protein tyrosine kinase gene family, of which *c-src, c-met, c-abl, c-erb-B2,* the EGF and PDGF receptors and others are members.

Isolation of new members of protein kinase and other gene families was carried out prior to the advent of PCR using degenerate oligo probes coding for conserved amino acid sequences.[115] PCR has enabled these studies to be extended considerably by the use of two degenerate oligos to amplify a DNA fragment of predictable size, encoding the active site of the catalytic domain. Wilks[116] used degenerate oligos encoding two highly conserved, adjacent amino acid sequences of the protein tyrosine kinase domains, PTK1 and PTK2, to amplify cDNA derived from a murine hemopoietic cell line. PCR products of 210 base pairs were produced, the subsequent sequencing of which revealed five sequences which differed in the nonconserved regions between the PTK-specified sequences. Two of these represented novel tyrosine kinase genes. In further studies, Wilks and colleagues[117] examined the parameters affecting efficiency of such simultaneous amplification of related gene sequences. Since the highest degree of conservation in the tyrosine kinase domain is confined to 3-5 amino acid stretches of 100% identity, wither lesser homologies to either side of these, short oligos with low degeneracy (14-17 nucleotides plus a 5' restriction enzyme site) corresponding to highly conserved regions were compared to longer but more degenerate oligos (24-26 nucleotides, plus restriction enzyme site), overlapping the same region. It was envisaged that the extra 5' sequences would facilitate amplification after the first copy of the template had been synthesized. At low annealing temperatures (~35° C; 2 min), both short and long oligos produced similar amounts of PCR products of the correct size. However, the efficiency of the longer oligos declined drastically at $T_{ann}=40°$ C, while the short oligos sustained a constant level of amplification up to $T_{ann}=50°$ C. A nonexhaustive analysis of the latter identified seven different tyrosine kinase gene sequences, of which two had been previ-

ously unknown. Thus in searching for novel members of gene families, a further source of degeneracy is encountered when designing oligos in that the amino acid sequence itself may be somewhat redundant. Better results are therefore obtained when highly conserved peptide sequences are selected, however short they may appear, down to a minimum of five amino acids. This will ensure the lowest possible degeneracy and enable annealing temperatures to be raised to obtain greater specificity for the gene family.

Similarly, the conserved amino acid motifs surrounding the variable region of the immunoglobulin heavy and light chain genes were used by Leboeuf et al[118] to characterize hybridoma antibodies. In this early study, specific cDNA synthesis was carried out using a single degenerate oligo to direct reverse transcription, taking advantage of immunoglobulin mRNA abundance. This was followed by PCR amplification using the same primer with another degenerate oligo coding for conserved sequences flanking the opposite end of the variable region.

The power of this method was demonstrated by Kamb et al[119] who used amino acid sequences from embryo pattern formation and tyrosine kinase gene families of the fruit fly, *Drosophila melanogaster,* to produce degenerate oligos which amplified similar cDNA sequences derived from the nematode, *Caenorhabditis elegans.* Genes encoding potassium channel proteins, involved in neuronal firing patterns, were identified in human genomic DNA using *D. melanogaster* and murine protein based PCR primers. Thus the use of degenerate oligos for PCR amplification not only extends to other members of gene families within one organism but can be used to identify evolutionarily conserved genes of organisms belonging to completely different phylogenetic groups.

APPLICATIONS OF DEGENERATE OLIGO PCR

Gene Cloning Via Peptide Sequence

Techniques enabling the sequencing of very small quantities of proteins have in recent years increased the demand for a quick and reliable method for isolating their encoding genes. Degenerate oligo PCR fulfils this requirement and has been applied to many areas of biological research.

A cDNA encoding a major human placental protein, tyrosine phosphatase, was isolated using cloned PCR products amplified by degenerate oligos based on the amino acid sequence of the isolated protein.[120] Diacylglycerol kinase, a modulator of protein kinase C in the calcium-dependent signal transduction pathway, was purified and partially sequenced, leading to cloning of its encoding cDNA through degenerate oligos encoding short stretches of the amino acid sequence.[121] The tumor necrosis factor receptor, expressed at low levels in a number of cells, was purified from the HL60 leukemic cell line and partially sequenced, leading to isolation of a coding cDNA.[122] Sequence homology to the NGF receptor suggested that these receptor proteins belonged to a new gene family. Similarly the cloning and sequencing of the cholecystokinin A receptor cDNA from rat pancreas via a peptide sequence showed that it belonged to the 'superfamily' of guanine nucleotide-binding, regulatory protein-coupled receptors.[123]

Synthesis of proteins using suitable vector-host systems to express full-length cDNAs has greatly facilitated the study of enzyme-catalysed biochemical pathways. The ubiquitin-dependent protein degradation pathway, involved in protein turnover and its regulation in the cell, utilizes the activity of an enzyme which catalyses the synthesis of multi-ubiquitin chains which 'flag' proteins destined for degradation by proteases, etc. The regulatory role of this protein may be more closely defined since the isolation of its cDNA through the use of degenerate oligo PCR.[124] The attachment of oligosaccharides to proteins is often vital to their correct function, but this phenomenon is as yet only partially understood. The elucidation of glycosylation pathways has been advanced in recent years by the isolation of enzymes responsible for the attachment or trimming of oligosaccharides, such as the endoplasmic reticulum alpha mannosidase of rat, for which the coding cDNA was cloned through degenerate oligo PCR amplification.[125]

The medicinal leech, *Haemanteria officinalis*, has recently attracted interest as a source of anticlotting agents. cDNA clones were isolated by Keller and coworkers[126] using degenerate PCR oligos based on partial amino acid sequences of leech antiplatelet protein. A recombinant protein expressed in yeast from one such cDNA was shown to inhibit collagen-stimulated platelet aggregation with similar efficacy to the native protein isolated from leech.

Fel-dI, the major allergen produced by the domestic cat, was purified by immunoaffinity from house dust and partially sequenced. It was found to consist of two protein chains, and cDNA derived from cat salivary gland was amplified using appropriate degenerate PCR primers, leading to identification of the full-length coding sequences.[127]

The human myeloid cell differentiation antigen is specifically expressed by, and may be isolated from, cells of the granulocyte and monocyte lineages and is induced by interferon alpha. Isolation of its cDNA via PCR amplification with degenerate oligos led to examination of the gene structure, revealing transcriptional regulatory elements which are known to be responsive to interferon alpha. Aspects of the full protein sequence suggest that the antigen is itself a regulator of gene transcription and may therefore be involved in early stages of the interferon alpha response of these cells.[128] cDNA encoding the soluble angiotensin binding protein was similarly cloned via knowledge of the amino acid sequence of selected tryptic peptides.[129]

Conservation across species, and even phylogenetic orders, of sequences specifying protein function was demonstrated by the cloning of a cDNA for a fatty acid binding protein (FABP) of locust flight muscle.[130] The translated protein sequence bore significant and widespread homology to previously characterized mammalian FABPs, particularly one specific to human heart.

Cloning Novel Members of Gene Families

The observation that functionally related proteins contained regions of conserved amino acid sequence led to the grouping of these into 'gene families'. Usually the regions of homology in these genes are too small, or the coding nucleotides too degenerate, for the practical identification of novel gene family members by conventional hybridization techniques. Degenerate oligo probes have been used with some success, but degenerate oligo PCR is now the method of choice in the analysis of gene families.

The major area of research in gene families is in signal transduction pathways, concentrating on the guanosine nucleotide binding proteins (G proteins), which frequently take the form of a signal receptor, and the protein kinases and phosphatases, which are involved in the regulation of proteins which make up the signalling pathways. A wide variety of genes belonging to the G protein 'superfamily' have been isolated by degenerate oligo PCR. A human thyrotropin receptor involved in thyroid autoimmune disease, expressed at very low levels on thyroid cells, was cloned via PCR with primers encoding conserved G protein sequences, using deoxyinosine at ambiguous nucleotide positions.[131] The canine histamine H2 receptor was cloned using G protein degenerate PCR oligos to yield an expressible cDNA which could be used in studies on the effects of the H2 antagonist, cimetidine.[132] A D1-dopamine receptor was cloned in this way from a neuroblastoma cell line and shown in tissue culture cells to be linked to the activation of adenylyl cyclase,[133] a gene which itself belongs to a separate gene family.[134] The A1 adenosine receptor was cloned through G protein encoding PCR oligos.[135] In a variation on the standard method of first cloning and characterizing PCR products, a rat hippocampus cDNA library was amplified by degenerate oligo PCR using primers directed at G protein conserved regions, and the resulting DNA fragments used as a single probe against a human hippocampus cDNA library. This led directly to the isolation of a human cDNA with high homology to the known rat alpha 1A adrenergic receptor.[136] The human cannabinoid receptor was isolated using G-protein specific degenerate primers and identified by high homology with the corresponding rat gene.[137] The bovine neuropeptide Y receptor

was identified and cloned by analysis of G protein degenerate oligo directed PCR products,[138] as was a human 5HT1-like serotonin receptor.[139] The technique clearly has the potential to isolate all G proteins of whichever subfamily is specified; thus a number of novel G proteins of unknown function have been isolated as well as those of specific interest.[140,141]

Protein tyrosine kinases and phosphatases comprise two other large gene families. PCR using degenerate oligos as described above[116] has been used to dramatically increase the list of known tyrosine kinase[116,117,142-144] and phosphatase[145,146] sequences, but not all of these have yet been assigned functions. The amount of work required to biochemically characterize these proteins dwarfs that involved in their initial isolation, a major problem in the analysis of gene families. However, homologies between other domains of the encoded protein and sequences of known function often provide clues as to the function of novel gene family members.[144]

Genes related to the noradrenaline and gamma-aminobutyric acid (GABA) transporters have been isolated using degenerate primers derived from the conserved regions of these proteins, indicating that they belong to a family of neuronal transporter proteins. These include rat serotonin[147] and L-proline transporters,[148] plus a spinal cord-specific neurotransmitter transporter of unidentified function.[149] A murine cyclic nucleotide phosphodiesterase was isolated, using degenerate oligo PCR, from a lymphoma cell line cDNA library.[150]

The major regulatory proteins involved in the development of complex organisms often, if not always, belong to gene families. The paradigm of these developmental regulators is the Hox gene family, all members of which contain a 'homoeobox' sequence which is conserved across highly extended evolutionary pathways. The degenerate oligo PCR method is ideally suited to the cloning of new members Hox and other developmental gene families,[151,152] the identification of these novel genes in relation to known family members in other species[153,154] and their characterization in terms of spatial and temporal distribution.[153,155]

Cloning InterSpecific Gene Homologues

The isolation of genes related by protein function also suggested that functionally conserved proteins could be identified in different organisms by amplifying cDNA with degenerate PCR primers encoding not only the amino acids observed in the conserved peptide domains of the 'reference' species, but also functionally related residues such as serine and threonine, or histidine and arginine. The ability to isolate gene homologues of this kind in different species is vital as it enables the researcher to identify a laboratory animal model for the study of topics of relevance to biological and medical research. Cloning the mouse homologue of an oncogene, for example, may lead to the engineering of a transgenic animal model to mimic the various stages of carcinogenesis, while the discovery of functional homologues of mammalian genes in more primitive organisms such as yeast and bacteria permits the study of these conserved proteins in a less complex environment.

The lit/lit mouse is a model for human growth hormone deficiency. The mouse homologue of the growth hormone-releasing hormone (GRH) was cloned by degenerate oligo PCR using primers derived from 6-amino acid long conserved regions of GRH peptides of human and rat.[156] The resulting GRH probe was used to demonstrate accumulation of GRH mRNA in the mouse model, suggesting a blockade downstream of this signal with a concomitant positive feedback acting on the transcriptional regulation of the GRH gene. Homologies between all known brain-derived neurotropic factors and nerve growth factors were used to construct degenerate primers to amplify rat hippocampal cDNA resulting in the isolation of a novel gene with significant regions of homology to both molecules, which expressed a protein with neurotropic activity.[157] The rat homologue of the murine thyrotropin-releasing hormone receptor was also cloned using degenerate oligos derived from regions of the murine protein sequence.[158] Working in the opposite direction, human genes for cardiac troponin I[159] and a brain glutamate receptor[160] were isolated by

homology to conserved regions of their homologues in other mammalian species, using degenerate oligos coding for these amino acid sequences.

The fruit fly, *Drosophila melanogaster*, is currently the most advanced model of eukaryotic development. A number of important developmental gene families discovered in Drosophila are highly conserved in higher eukaryotes, the most notable example being the Hox gene family.

Some genes are not as extensively conserved at the amino acid level as their functional conservation might suggest. The isolation of related genes by PCR using degenerate oligos encoding the few commonly retained short sequences of amino acids provides a significant advance in such cases over conventional oligo probing of cDNA libraries, in which the necessary constraints on hybridization stringency lead to a high degree of nonspecificity. The wide ranging study of Kamb et al[119] exemplifies this approach. In addition to these studies, Drosophila and mouse potassium channel protein sequences were also used to derive primers for conserved sequences to isolate a rat cDNA homologue.[161] The proto-oncogene *c-rel* shares a conserved domain with the DNA-binding subunit of nuclear factor kappa B and the Drosophila morphogen, 'dorsal', a homology which was used to isolate a cDNA encoding human subunit of the NF kappa B transcription complex with DNA binding activity.[162]

The identification of cell cycle gene homologues in two highly divergent yeast species (*S. pombe* and *S. cerevisiae*) and mammals has lent considerable importance to the study of cell division control in yeast. Conservation of cyclin B in a range of organisms from humans to fission yeast (*S. pombe*) led to the isolation of its gene homologue in *S. cerevisiae*, enabling experiments to be carried out with cyclin B mutants which demonstrated both its partial functional redundancy and a potential role in cell cycle arrest.[163] The murine cdc25 cell cycle control gene was cloned using degenerate oligos based on the conserved sequences of cdc25-related proteins of yeast.[164] Functional complementarity was shown between yeast protein and its conserved mammalian homologue by substituting conserved domains in each cDNA and demonstrating genetic suppression of loss of function mutants by the translated chimeric protein.

Despite extensive conservation of functional regions, it may not always be the case that a gene which has homologues in yeast and mammals is essential to fundamental life processes. Thus cyclin B was found to be partially functionally redundant, and a gene isolated from yeast using degenerate oligos for mammalian serine/threonine phosphatases was found to be nonessential for yeast cell growth.[165] It may be that protein functions, like sense organs and limbs, are duplicated to reduce the evolutionary disadvantage of losing one gene or adapting it to another purpose.

The cloning of functional homologues has been extended across phyla and even into the plant kingdom. Cloning via degenerate oligo PCR of carrot genes containing characteristic conserved regions of the proliferating cell nuclear antigen, involved in normal DNA replication in mammalian cells, revealed a remarkable degree of homology, even to the level of intron positioning.[166] An alternative strategy was employed by Feiler and Jacobs[167] in searching for the pea equivalent of cdc2, an essential cell cycle control protein conserved from yeast to man. The reconstitution of immunoreactivity in the translated protein by the replacement of a conserved region of the cloned cdc2 cDNA by PCR products generated from peas using degenerate oligos, identified sequences derived from the pea cdc2 homologue.

The high degree of conservation of RNA polymerase amino acid sequences was exploited by Engel et al[168] to amplify homologous sequences from *Chlamydia trachomatis*. Expression of these sequences was demonstrated early in *C. trachomatis* infections, and the isolation of such genes could have implications for the development of new drugs specifically targeted at fundamental biochemical processes of infectious agents. For instance, cryptosporidiosis does not respond to antifolate drugs which are effective against parasites closely related to the pathogen (*C. parvum*). In protozoa, dihydrofolate reductase (DHFR) is encoded by a bifunctional gene which also encodes thymidylate synthase (TS).

The DHFR-TS gene of *C. parvum* was therefore cloned using degenerate oligos based on other TS sequences and shown to be most similar to that of the malarial parasite, Plasmodium.[169] The new sequence could be used to identify new therapies for cryptosporidiosis. Similarly, a number of members of the serine protease gene family were identified in a parasitic nematode, *Anisakis simplex*, by degenerate oligo PCR based on highly conserved eukaryotic serine protease peptide sequences.[170] As an initial step in vaccine development, the thymidine kinase gene of feline herpesvirus 1 was isolated by taking advantage of the conserved regions of this gene in the herpes virus family to construct degenerate oligos.[171] A similar approach has also been used in identifying the thymidine kinase gene of the infectious laryngotracheitis virus, based on homologous regions of the ATP and thymidine binding sites of other herpesviruses.[172] The DNA polymerase gene of human herpesvirus 6 was also identified and cloned in this way.[173] Degenerate oligos to conserved genomic regions of large virus families can also be used in attempts to isolate novel virus types.[174]

The Use of Degenerate Oligo PCR as a Diagnostic Aid

Many disease-causing viruses are grouped into genetically related groups which share extensive homologies at the protein level. Examples of these are the papillomaviruses and herpesviruses. Protein sequence similarity may also occur between infectious protozoa or bacteria of the same genus. Despite amino acid sequence homology, however, codon usage may be highly redundant due to evolutionary alterations at sequence level which do not ultimately affect the translated protein sequence. This results in considerably lower sequence matching at the genome level. Any general diagnostic application screening for the presence of DNA of such a pathogen 'family' therefore requires a large number of probes (or PCR primers) to allow for all known sequence variations, which may even so fail to identify previously unknown infectious agents of the same related group. DNA diagnosis therefore becomes laborious, slow and expensive. Moreover, the levels of pathogen DNA in a sample may be too low to detect by conventional hybridization techniques.

PCR using degenerate primers designed to encode related amino acid sequences has been used in a diagnostic context as a rapid identification procedure. The technique was applied by Snijders et al[175] as a general method for detection of human papillomavirus infection, using primers based on the partially conserved major viral capsid protein, L1. A two-step PCR method, using nested degenerate primer sets based on papillomavirus L1 sequences, has been used successfully to detect low-level human papillomavirus DNA sequences in patients.[176-178] Detection of papillomavirus oncogenes in head and neck carcinoma samples using degenerate PCR oligos encoding conserved regions of oncogenic viral subtypes was supported by amplification of L1 genes followed by specific restriction endonuclease digestion and gel electrophoresis of the resulting PCR products to identify virus type.[179]

Members of the dengue virus family vary both in their genomic composition and pathology of infection. A preliminary general diagnostic system for dengue virus infection was devised by Henchal and colleagues,[180] based on amplification of sample material by degenerate prime sets. The resulting amplified DNA could then be brought forward into a specific test based on hybridization to known viral DNA probes derived from the genomes of pathogenic dengue virus types.

Conservation of the major outer membrane protein amino acid sequence of Chlamydia enables the simultaneous amplification of DNA of the encoding gene, *ompA*, from more than one species, using degenerate oligo PCR. Kaltenboeck and coworkers[181] utilized this observation to devise a diagnostic test. The *ompA* gene was amplified from sample material using oligos containing deoxyinosine at positions of ambiguity. The amplified product was then subjected to a second round of PCR, this time using species-specific, nondegenerate primers annealing to sequences internal to the original oligos. PCR product length and restriction endonuclease digestion fragments could then be used to determine the Chlamydia species of origin.

Subspecies and strains of *Onchocerca volvulus* are characterized by variability in a 150 nucleotide tandemly repeated sequence, which can be detected using specific DNA probes. Difficulty in obtaining hybridization specificity and in isolating genomic DNA from larvae and skin microfilariae of the parasite led to the development of degenerate PCR oligos to amplify these samples, producing DNA fragments which could be identified by the probes.[182]

Degenerate oligo PCR, as a rapid assay, may therefore solve some of the problems posed by the genetic variation of pathogens, a major limiting factor in making PCR a cost-effective diagnostic technique as the use of large numbers of specific oligos represents a large initial outlay and involves a great deal of operator time in terms of repeating tests with new sets of oligos.

SEQUENCING PCR PRODUCTS

If PCR[183] was the greatest biotechnological advance of the 1980s, the 'dideoxy' method of sequencing DNA[184] was that of the 1970s and one of the two major enabling factors in the development of PCR. In turn, new PCR techniques have dramatically shortened the timescale of the sequencing method by providing a partially purified sequencing template in ample quantity. Further refinements and adaptations have speeded up the process even further until it has become possible to obtain nucleotide sequences within 24 hours of obtaining a DNA sample.

Dideoxy sequencing requires a single-stranded DNA template, which is conventionally provided by the M13 bacteriophage cloning vector. DNA for sequencing must therefore first be subcloned into M13, which is then purified from liquid culture, and this is the rate-limiting step in gene sequencing. The slow process of obtaining specific clones for sequencing limited large-scale sequencing projects such as those aimed at assessing sequence variations of a particular gene within a population, or associated with heritable disease. In such cases, if a gene sequence is known, PCR primers give direct access to specific DNA (or mRNA) sequence segments in genomic or mRNA isolates without the need to construct, screen and subclone genomic or cDNA libraries. Simple techniques enable the double-stranded PCR product to be sequenced as single-stranded DNA. PCR can also be used to quickly acquire sequence data from plasmid or phage vector recombinants directly from screened DNA libraries, as single bacterial colonies or phage plaques may be amplified directly, without purification of DNA, using primers flanking the cloning site. PCR is therefore ideally suited to the automation of template preparation in large scale sequencing projects, as bacterial colonies grown on grids can be transferred directly into microtiter plate wells containing reaction mix.

Direct sequencing of PCR products also avoids one of the problems associated with the cloning of PCR products: the occasional random misincorporation of nucleotides by Taq polymerase, which tends to occur more frequently as reaction conditions become limiting. If PCR products are sequenced en masse, however, any random mismatches will not be detectable against the much stronger signal given by the correct sequence.

Over the past few years, a number of sequencing methods applicable to PCR products have been developed. This chapter describes those presently in common use.

PCR SEQUENCING METHODS

Sequencing of Double-Stranded Templates

Around the time when the polymerase chain reaction was first developed,[183] sequencing of double-stranded, supercoiled plasmid DNA was commonly carried out by heat or alkali denaturation followed by annealing (or neutralization) in the presence of a specific sequencing primer. This method was convenient for obtaining limited sequence from relatively crude DNA preparations and useful in reducing the work of subcloning large numbers of DNA inserts into M13 phage vector. It could also be readily applied to the sequencing of PCR products which were first purified by gel electrophoresis, DNA affinity column purification or even simple isopropanol precipitation. Thus early reports on PCR sequencing utilized this direct approach, either

with conventional sequencing reagents or with Taq polymerase and adapted dideoxy-nucleotide/dNTP ratios.[185-187] Taq polymerase was used as the sequencing enzyme as it was considered that DNA template secondary structure would be less likely to form and impede enzyme processivity at the enzyme's optimum operating temperature of 72-75° C. This proved to be a straightforward and very rapid method for obtaining sequence data directly from genomic DNA or the cDNA products of reverse transcription of RNA. As a result, several commercial kits have become available which are specifically designed for high temperature sequencing of double-stranded templates. However, the method has its limitations in terms of the length of sequence which can be obtained, which is generally limited to around 150 nucleotides. Short, linear molecules such as those produced by PCR are difficult to irreversibly denature in the same way as a supercoiled plasmid; rapid reassociation of complementary sequences occurs at conventional sequencing temperatures due to the mobility of the denatured strands, affecting enzyme processivity. When Taq polymerase is used to maintain a higher sequencing reaction temperature, reduced concentrations of dNTPs appear to increase the rate at which the enzyme 'falls off' the template strand, resulting in false terminations.[188] This may be the reason why some researchers have difficulty in using this method reproducibly, and turn to other PCR sequencing methods which have been developed with the aim of eliminating sequencing artifacts.

Sequencing of Single-Stranded DNA Produced by Asymmetric PCR

The dideoxy sequencing reaction works optimally on purified, single-stranded DNA templates from which 500 nucleotides or more of DNA sequence may be obtained routinely. The speed with which sequenceable DNA is generated by PCR is mitigated by the short (150 nucleotide) sequences which are obtained when double-stranded reaction products are sequenced, and by the tendency of this method to produce artifacts. Although this may not be a problem in situations where short sequences are screened, e.g., codon 61 of the *ras* oncogene, if larger sequences are to be analyzed, e.g., exons 4-8 of the p53 gene, a long-range PCR sequencing method is required.

Single-stranded DNA can be produced in a thermal cycling reaction using asymmetric PCR.[188-190] In this reaction, template DNA is amplified according to the precalibrated cycle parameters (Chapter 3), except that one of the oligo concentrations is reduced to one-tenth of the normal level. The limiting oligo must be that which primes the DNA strand from which sequence is required. This oligo is quickly used up as the reaction progresses, producing a relatively small amount of double-stranded DNA. The other oligo is still present in excess, however, and continues to prime new DNA strand synthesis (Fig. 7). These new DNA strands now have no complementary sequences with which to anneal, and

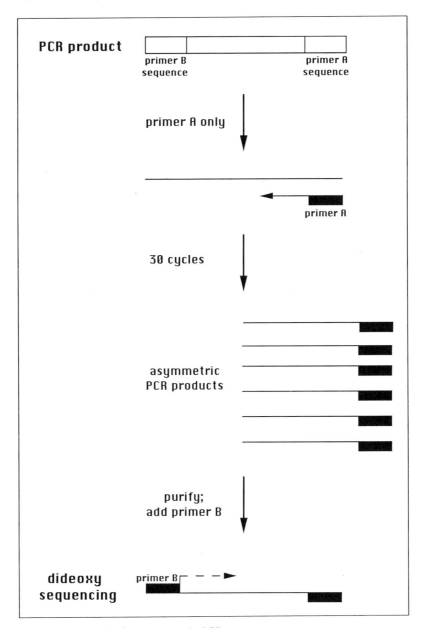

Fig. 7. Sequencing by asymmetric PCR.

therefore accumulate arithmetically as the existing, double-stranded PCR products are reprimed on one strand during each cycle. Over a PCR run of 40 to 50 cycles, these single-stranded molecules become the major reaction product, at a 10- to 20-fold excess over double-stranded DNA. This single-stranded DNA becomes the template for the sequencing reaction.

Although the small amount of double-stranded PCR products can be ignored for the purposes of sequencing, there still remains in the reaction solution a sizeable quantity of unincorporated nucleotides and primers. These must be removed as they are capable of taking part in the sequencing reaction. Repeated precipitation in 2M ammonium chloride, 50% isopropanol, followed by a careful 70% ethanol wash, is effective in both concentrating the DNA and removing PCR reagents.[190] More stringent measures are taken by some researchers, however, such as centrifugation and washing in a microconcentrator.[189,191] One problem associated with precipitation is that the small DNA pellets are easily dislodged and lost: the addition of 1 μl of purified glycogen before precipitation both increases DNA yield and gives a visible pellet which sticks to the bottom of the microcentrifuge tube.

Products of asymmetric PCR may be sequenced using standard techniques, except that templates should be subjected to only moderate denaturation (65° C for 2 minutes) before primer annealing to prevent the residual double-stranded DNA from taking part in the reaction.

Using this method of generating single-stranded sequencing templates results in sequences of clarity and length which is comparable with M13 phage sequencing. However, asymmetric PCR suffers from two disadvantages. The first is that the ratio of primers added at the beginning of the reaction may have to be calibrated for different sequences in order to obtain a sufficient yield of single-stranded DNA. The second is that there is no simple and rapid means to quantify the end product. Single-stranded DNA absorbs about ten times less ethidium bromide in agarose gels than the corresponding amount of double-stranded DNA and does

not always appear as a single band. Quantitation therefore needs to be carried out by labelling pilot reactions by the addition of ~1 μCi of ^{32}P-dCTP,[190] separating products by gel electrophoresis and analyzing an autoradiograph of the dried-down gel by scanning densitometry. The signal obtained from single-stranded products is compared to known radioactive standards which are included in the same gel. Such is the extent of work required for this exercise that it would be simpler in the long run just to perform sequencing reactions on aliquots of a series of asymmetric PCRs containing various ratios of input oligos, and select the best sequencing readout.

Alternatively, single-stranded DNA yield can be guaranteed by purifying a known quantity (~0.5 picomoles) of double-stranded PCR product and subjecting it to 30 thermal cycles in a PCR mix containing only one oligo.[191] The single-stranded product of this reaction is purified and sequenced in the same way as described above. Extensive purification of the original PCR product may not be necessary; indeed, the authors have found that, with a clean amplification, simple isopropanol precipitation and 70% ethanol washing will often suffice. The only disadvantage of this modification of the asymmetric PCR technique is the extra time taken to produce the double-stranded PCR product, quantify it by comparison to known standards on an agarose gel and carry out a second PCR run.

Recently, however, even faster PCR sequencing methods have been developed.

Sequencing on a Solid Support

Clear, artifact-free sequencing depends on the separation of the single-stranded DNA template from any molecules which may interfere with the sequencing reaction in the first instance, and the separation of unincorporated nucleotides, radiolabel, enzyme protein and glycerol from the sequencing reaction products after the reaction has taken place. A number of techniques are employed in both conventional and PCR sequencing to obtain purified template DNA, while electrophoretic fractionation on a denaturing polyacrylamide gel is relied upon to separate nonspecific signals from true sequence.

A novel approach to sequencing PCR products, which yields remarkably clean sequencing gels, entails the capture of the DNA strand to be sequenced on a solid support. Annealing of the sequencing primer, the sequencing reaction and removal of unincorporated reagents is carried out on the immobilized DNA strand, allowing the final release of highly purified sequence fragments. To make this possible, an 'anchor' must be incorporated into the sequencing template during amplification. The anchor takes the form of a chemical modification which distinguishes the sequencing template from its complementary strand and is recognized and bound tightly by the solid support. Furthermore, the modification must in no way interfere with the sequencing reaction.

Biotin is the compound of choice for use as an anchor for PCR-generated DNA strands and is chemically linked to the primer directing synthesis of the required sequencing template, thus distinguishing this strand from its complement. Biotin is bound tightly by streptavidin, which may in turn be linked to a number of solid matrices. The solid support may take the form of the plastic base of each well of a 96-well microtiter plate. If an automated sequencer is available, an entire PCR run and sequencing reaction can take place in one microtiter well for each sample. In most research laboratories, however, there is no requirement at the present time for sequencing on such a grand scale, and the technology required for automated sequencing represents a very large initial outlay and must be run continuously to be cost-effective. In the conventional sequencing reaction, each sample requires one reaction vessel for each of the four nucleotides, A, C, G and T, so the use of streptavidin-coated microtiter plates would lead to considerable wastage.

Recently, paramagnetic particles coated with streptavidin have been developed for sequencing and other molecular biological procedures.[192] These are tiny magnetizable beads which can be kept in suspension by gentle agitation. Moving the reaction vessel close to a high flux density magnet, however, results in the adherence of the beads to the side of the tube nearest the magnet. The beads can be readily resuspended after removal of the magnet, retaining no inherent magnetic properties. PCR products synthesized with one biotinylated primer are incubated with a suspension of the beads in a microcentrifuge tube and bind strongly. The suspension is then placed next to a high flux density magnet and the beads, together with the bound DNA, migrate to that side of the tube. The liquid component may now be removed without significant loss of beads, which may be washed with distilled water to remove any traces of the original PCR mix. PCR products are thus purified to a high degree by a procedure which requires less than five minutes of hands-on time. The non-biotinylated strands are then dissociated from the immobilized DNA by alkaline denaturation and may be completely removed by remagnetizing the beads and withdrawing the unbound DNA-containing liquid.[193-195] Highly purified, single-stranded DNA templates are all that remain on the beads. The amount of paramagnetic beads used is small enough to be resuspended in the very small volumes required for the sequencing reaction. Sequencing primer annealing and the reaction itself are carried out with the bead suspension; once the reaction is complete, unincorporated reagents are removed in the liquid phase and the beads resuspended in denaturing sequencing gel loading buffer. Sequencing reaction products are then liberated by heat denaturation and the immobilized template strand effectively removed by a further round of magnetization. The liquid phase now contains the radiolabelled sequence fragments which are loaded directly onto a sequencing gel. The resulting sequence autoradiograph is strikingly clean and free from artifacts, and the problem of disposal of the large volumes of radioactive electrophoresis buffer produced by running off unincorporated radiolabelled nucleotide in conventional sequencing is avoided. The only potential technical disadvantage of this method lies in the signal strength from the gel, which tends to be four or five times lower than that obtained with conventional sequencing. This is probably due to settling of the paramagnetic beads which occurs during the sequencing reaction, with consequent microenvironmental reagent limitation, and could be avoided by gentle rotation of the reaction tubes during incubation. This technique is also expensive, both in terms of initial outlay on the high flux density magnet, and in ongoing costs of biotinylated oligos and streptavidin-coated paramagnetic beads.

Sequencing of PCR Products Via Selective Exonuclease Digestion

The main concern when sequencing products of the polymerase chain reaction is to displace or remove DNA strands which are complementary to the sequencing template and which are able to interfere with the sequencing reaction via reassociation with the template strand. Asymmetric PCR and solid support sequencing were developed to obviate this problem, but the former involves extra steps which are time-consuming, and although paramagnetic bead technology represents a significant advance, it is a very expensive procedure.

An additional template preparation technique, which has recently become popular, entails the removal of the nontemplate strand by enzymatic digestion.[196] Gene 6 of the T7 bacteriophage encodes an exonuclease enzyme which specifically digests one strand of a double-stranded DNA molecule, progressively removing nucleotides from the 5' end of the strand.[197] The enzyme is commercially available and commonly used for preparing single-stranded DNA template for sequencing. Digestion of PCR products with the enzyme proceeds from the 5' end of each DNA strand, i.e., from both ends of the molecule (Fig. 8). The two exonuclease enzymes carrying out this process thus move towards one another along the DNA fragment. When they meet, the remaining undigested DNA consists of two single strands of approximately equal length, representing two halves of the original fragment on alternate strands. These single strands have no complementary DNA as this has been digested away by the enzyme, and because the digested DNA was removed in a 5' to 3'

direction, they each contain a binding site at the 3' end for one or other of the original PCR oligos. Therefore, the addition of one of these oligos to the completed digest supplies primed sequencing template.

For exonuclease digest and sequencing, 0.5 to 1 picomole of PCR-amplified DNA (isolated from agarose gel, column-purified or simply isopropanol precipitated with 70% ethanol wash) is added to the conventional sequencing reaction buffer with 5 units of T7 gene 6 exonuclease and subjected to a 10 minute digest at 37°C. The enzyme must be inactivated before sequencing can be carried out, and 15 minutes at 80°C is sufficient to destroy all exonuclease activity. Thus, single-stranded DNA of high enough purity for sequencing may be produced within half an hour of obtaining double-stranded PCR products. Sequence lengths comparable to those of conventional M13 sequencing are easily obtainable (assuming the PCR product is long enough in the first place).

Occasionally, using the original PCR oligos as sequencing primers fails to yield good results, and new primers annealing a short distance within the template sequence may be required.[198] As already mentioned, the exonuclease activity meets in the middle of the double-stranded DNA fragment, effectively producing two 'half sequences'. If the full sequence of an amplified DNA fragment is to be scanned, the 'overlap' region in the middle may not give a strong signal and, moreover, the method only allows sequencing in one direction for each of the single-stranded fragments. It is, however, also possible to obtain full-length single-stranded template for each DNA strand of a PCR product by exploiting the inability of T7 gene 6 exonuclease to digest single-stranded DNA. Certain restriction enzymes, e.g., PvuI, digest DNA to leave a short, single-stranded 'overhang' at the 3' ends. Incorporation of two such unique restriction enzyme recognition sites into the ends of the amplified DNA can be carried out by adding the appropriate nucleotide sequences to the 5' ends of the PCR primers. Digesting the amplification products obtained with these primers with one or other restriction enzyme will produce

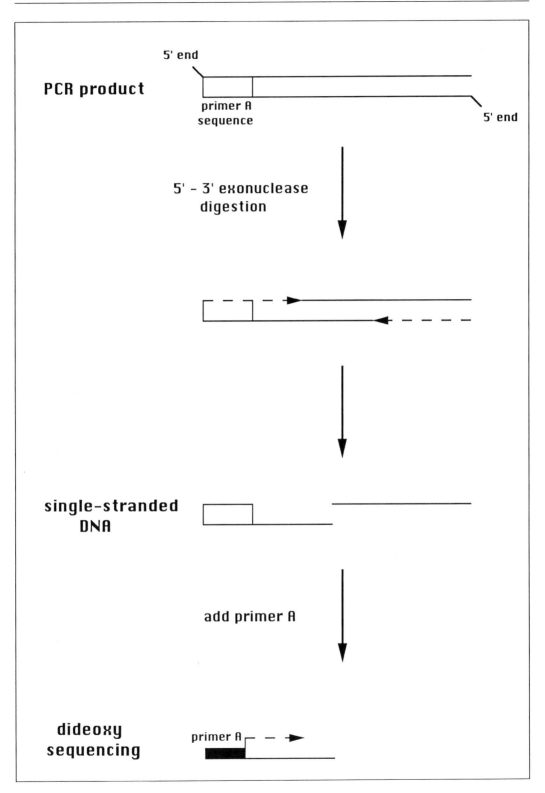

Fig. 8. Sequencing exonuclease digested PCR products.

a 3' overhang at one end of the DNA, preventing T7 gene 6 exonuclease digestion of that DNA strand. Cutting with the other restriction enzyme will protect the other strand. It is possible to carry out restriction digests, followed by exonuclease digestion in the sequencing buffer to save time, but it is not uncommon for restriction enzyme digestion of sites located at the ends of DNA molecules to proceed inefficiently even under ideal conditions. It is therefore advisable to carry out digests in appropriate reaction conditions, followed by column purification or ethanol precipitation, and to incorporate two or three extra nucleotides at the 5' ends of the PCR oligos.

'Cycle Sequencing' Direct from Plaques and Colonies

Of the many rapid sequencing techniques which may be applied to PCR products, most are designed to accelerate the acquisition of sequence data once PCR products have been synthesized. One method, however, actually utilizes an amplification reaction to generate sequence from very small amounts of DNA such as are found in a single, plasmid-transformed bacterial colony or bacteriophage plaque.[199-201]

Phage or bacteria are lysed by a rapid procedure and separated into four aliquots, each of which is added to one of four reaction mixes. Each reaction mix contains radiolabelled sequencing primer, a dideoxy-/deoxynucleotide ratio precalibrated to give sequence information on each of the four DNA bases, and Taq polymerase in a suitable buffer system. The reaction tubes are then subjected to 20–30 cycles on a PCR machine. During each cycle, the DNA template is denatured and the radiolabelled primer anneals and is extended by Taq polymerase until a dideoxynucleotide is incorporated. Each new DNA strand is thus rendered incapable of further extension, in the same way as the conventional dideoxy sequencing reaction.[184] Over the 20–30 cycles, the concentration of these fragments builds up, and there is a tendency for longer strands to be synthesized as the concentration of the dideoxynucleotide strand-terminators de-

clines. By the end of the run, there are enough conventional sequencing fragments to produce a signal on a sequencing gel. The use of ^{32}P to label the sequencing primer ensures a relatively strong signal for the small amount of DNA synthesized. Because only the primer is labelled, the strength of the signal will be proportional to the number of fragments present. Longer sequences may be obtained by allowing the reactions to progress further, allowing terminations at greater distances from the primer.

This method, once calibrated, could provide an extremely rapid screen for large numbers of plasmid or bacteriophage clones containing unknown sequences, or to check the ligation junctions of subcloned sequences. However, as it appears to be difficult to obtain high quality sequence autoradiographs, the technique cannot yet be fully relied upon as a sequence data source.

CLONING PCR PRODUCTS

'Cloning' of DNA sequences entails the ligation of the free ends of a nucleic acid fragment to those of a linearized plasmid vector in such a way as to create a closed circle of DNA. This recombinant plasmid will self-replicate in bacteria, providing a means of generating large-scale quantities of any sequence of interest.[14]

DNA cloning presents a number of technical problems, one of which concerns the matching of ends of DNA fragments for ligation. Frequently, the desired insert sequence will be in the form of a restriction fragment with 'sticky ends' which are incompatible with those of the linearized plasmid. These can be converted into 'blunt ends', in which the ends of both DNA strands are flush. Double-stranded cDNA is also blunt-ended. Ligation of blunt ends, however, proceeds inefficiently and is non-directional, i.e., the insert may clone in either orientation in the vector.

The advent of PCR[184] meant that any sequence could be cloned into a plasmid or phage vector by ligation of PCR products. A number of groups utilized the new technique to adapt the ends of DNA sequences to enable directional cloning of PCR fragments.[202-207]

Using similar methods, it is also possible to influence eukaryotic expression of cloned sequences.

SYNTHESIS OF TERMINAL RESTRICTION ENZYME RECOGNITION SITES

In priming Taq polymerase activity, it is the nucleotide sequence of the 3' end of a PCR oligo which is most important. Mismatches at the extreme 5' end may affect the efficiency of amplification but generally do not affect specificity. It is therefore permissible to incorporate extra, nonmatching nucleotides at the 5' end of a specific oligo sequence to form a restriction endonuclease recognition site. The restriction site sequence will not anneal to the template DNA but, by the second cycle of a PCR run, a complementary copy of the whole oligo will have been synthesized and will become the major template sequence in the reaction (Fig. 9). Each oligo can be designed with different 5'

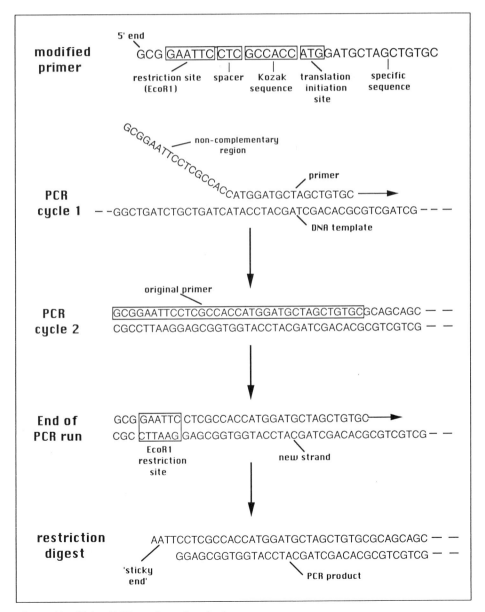

Fig. 9. Modifying PCR products for cloning

restriction site, so that the final PCR product will contain the specified restriction sites at either end. It may then be digested with those restriction enzymes to produce a sticky end which can be ligated into the cloning site of a plasmid or phage vector. The restriction sites may be placed in such a way as to specify the eventual orientation of the insert in the vector. This can be important for sequencing, expression of cDNAs or for restriction mapping and subcloning.

In designing oligos with 5' restriction sites, it should be borne in mind that most restriction enzymes are highly inefficient at digesting restriction sites at the extreme ends of DNA fragments. This is thought to be a result of the tendency of fragment ends to denature slightly under conditions favoring restriction enzyme activity. Digestion efficiency is markedly improved if two or three more nucleotides are added 5' to the restriction site of the oligo sequence. Usually a combination of G and C nucleotides is used, as these form stronger interstrand bonds (Fig. 9).

'T-A' CLONING SYSTEMS

Taq polymerase attaches an extra adenosine nucleotide onto the 3' ends of a PCR product. The 'A' nucleotide forms a 3' overhang with no complementary 'T' on the other strand.[208-210] This phenomenon has been utilized to improve cloning efficiency of PCR products, using a linearized vector with a 3' overhanging 'T' nucleotide.[211,212] The 'T' forms a sticky end of only one nucleotide, which nevertheless improves ligation and therefore cloning efficiencies. The only drawback of this system is that directional cloning cannot be achieved, although the vector may have a multiple cloning site which can be used for directional subcloning.

IMPROVING GENE TRANSLATION EFFICIENCY

PCR can be used to subclone coding sequences for expression in eukaryotic cells. Conventionally, this is carried out by directional subcloning into a plasmid expression vector containing appropriate transcriptional promoter and polyadenylation signals which ensure a high rate of synthesis of messenger RNA. The translation efficiency of the message may be increased

by altering the nucleotide sequence around the translation initiation signal (ATG) to the Kozak consensus sequence[213] (Fig. 9). PCR can be used to achieve this by incorporating the Kozak sequence into a PCR primer annealing to the 5' end of the coding sequence. A restriction enzyme recognition site can also be added to the 5' end of each primer to ensure directional cloning.

A CAUTIONARY NOTE:
TAQ POLYMERASE ERROR RATES

Any attempt to clone a gene using Taq polymerase amplification is subject to one important caveat. Under currently used PCR conditions, the enzyme occasionally incorporates an inappropriate nucleotide in the course of synthesizing a new DNA strand.[184] The frequency of such copying errors appears to be far higher than that of the low temperature DNA polymerases and may be a consequence of limiting conditions in the polymerase chain reaction. The absence of any 3'–5' exonuclease activity, the hallmark of so-called 'proof-reading' DNA polymerases, prevents Taq polymerase from correcting these errors, which are subsequently reproduced faithfully in the cloned PCR product, thus misrepresenting the true gene sequence.

The actual error rate of Taq polymerase has been a subject of some debate. Reports vary, but although rates as low as 6.6 per 100,000 nucleotides have been claimed,[214,215] it seems to be generally accepted that a misincorporation is likely to occur every 300-1000 nucleotides.[184] Variability in error rate estimates suggest that reaction conditions influence the likelihood of nucleotide misincorporation, with the majority of errors occurring in the plateau phase (Chapter 3: Problems and Pitfalls). This conjecture is supported by the observation that the same misincorporation is rarely found in more than one clone, suggesting that errors occur too late in the reaction to be significantly amplified. Certainly, reports claiming low error rates used low cycle numbers, but these groups also used lower dNTP and Magnesium concentrations[214,215] (see Chapter 2, Table2).

Misincorporations will occur in any PCR, so if gene cloning is the ultimate objective, at least 10 clones should be fully sequenced to obtain a consensus sequence for the gene.

Fortunately, there is little evidence that Taq polymerase targets particular sites for misincorporation (however, see ref. 216), so it is unlikely that any one nucleotide position will be invariably misrepresented.

Recently, two new thermostable DNA polymerases with proofreading properties, 'Vent' and 'Pfu' polymerase,[217] have become available. Although these two enzymes produce faithful copies of their templates, they are highly inefficient in the polymerase chain reaction, requiring a very high input of template DNA. This may not always be possible when cloning directly from tissue samples, for example, so a combination of low cycle number Taq polymerase amplification followed by high cycle number Vent or Pfu PCR may be required for faithful sequence copying.

REFERENCES

1. Larzul D, Guigue F, Sninsky JJ, Mack DH, Brechot C, Guesdon J-L. Detection of hepatitis B virus sequences in serum by using in vitro enzymatic amplification. J Virol Methods 1988; 20:227-237.
2. Gilliland G, Perrin S, Bunn HF. Competitive PCR for quantitation of mRNA. In: Innis MA, Gelfand DH, Sninsky JJ, White T, eds. PCR Protocols. San Diego, California: Academic Press Inc, 1990:60-69.
3. Becker-Andre M, Hahlbrock K. Absolute mRNA quantification using the polymerase chain reaction. A novel approach by a PCR aided transcript titration assay (PATTY). Nucleic Acids Res 1989; 17:9437-9447.
4. Pieper RO, Futscher BW, Dong Q, Ellis TM, Erickson LC. Comparison of O-6-methylguanine DNA methyltransferase (MGMT) mRNA levels in Mer⁺ and Mer⁻ human tumor cell lines containing the MGMT gene by the polymerase chain reaction technique. Cancer Communications 1990; 2:13-20.
5. Richardson CC. Bacteriophage T4 polynucleotide kinase. In: Boyer PD, ed. The Enzymes, 3rd Edition. New York, New York: Academic Press, Inc., 1981; 14:299.
6. Noonan KE, Beck C, Holzmayer TA, Chin JE, Wunder JS, Andrulis IL, Gazdar AF, Willman CL, Griffith B, Von Hoff DD et al. Quantitative analysis of MDR1 (multidrug resistance) gene expression in human tumors by the polymerase chain reaction. Proc Natl Acad Sci (USA) 1990; 87:7160-7164.
7. Gilliland G, Perrin S, Blanchard K, Bunn HF. Analysis of cytokine mRNA and DNA: Detection and quantitation by competitive polymerase chain reaction. Proc Natl Acad Sci (USA) 1990; 87:2725-2729.
8. Bettens F, Pichler WJ, de Weck AL. Incorporation of biotinylated nucleotides for the quantification of PCR amplified HIV-1 DNA by chemiluminescence. Eur J Clin Chem Clin Biochem 1991; 29:685-688.
9. Chelly J, Kaplan JC, Maire P, Gautron S, Kahn A. Transcription of the dystrophin gene in human muscle and non-muscle tissues. Nature 1988; 333:858-860.
10. Chelly J, Montarras D, Pinset C, Berwald-Netter Y, Kaplan JC, Kahn A. Quantitative estimation of minor mRNAs by cDNA polymerase chain reaction. Application to dystrophin mRNA in cultured myogenic and brain cells. Eur J Biochem 1990; 187:691-698.
11. Murphy LD, Herzog CE, Rudick JB, Fojo AT, Bates SE. Use of the polymerase chain reaction in the quantitation of mdr-1 gene expression. Biochemistry 1990; 29:10351-10356.
12. Kellogg DE, Sninsky JJ, Kwok S. Quantitation of HIV-1 proviral DNA relative to cellular DNA by the polymerase chain reaction. Anal Biochem 1990; 189:202-208.
13. Ballagi-Pordani A, Ballagi-Pordany A, Funa K. Quantitative determination of mRNA phenotypes by the polymerase chain reaction. Anal Biochem 1991; 196:89-94.
14. Sambrook J, Fritsch EF, Maniatis T. Molecular cloning: A laboratory manual. USA: Cold Spring Harbor Laboratory Press 1989.
15. Luqmani YA, Graham M, Coombes RC. Expression of basic fibroblast growth factor, FGFR1, FGFR2 in normal and malignant human breast, and comparison with other normal tissues. Br J Cancer 1992; 66:273-280.
16. Rappolee DA, Mark D, Banda MJ, Werb Z. Wound macrophages express TGF-alpha and other growth factors in vivo: Analysis by mRNA phenotyping. Science 1988; 241:708-712.

17. Hart C, Chang SY, Kwok S, Sninsky J, Ou CY, Schochetman G. A replication deficient HIV-1 DNA used for quantitation of the polymerase chain reaction (PCR). Nucleic Acid Res 1990; 18:4029-4030.

18. Lin JH, Grandchamp B, Abraham NG. Quantitation of human erythroid-specific porphobilinogen deaminase mRNA by the polymerase chain reaction. Exp Hematol 1991; 19:817-822.

19. Perrin S, Gilliland G. Site-specific mutagenesis using asymmetric polymerase chain reaction and a single mutant primer. Nucleic Acids Res 1990; 18:7433-7438.

20. Wang AM, Doyle MV, Mark DF. Quantitation of mRNA by the polymerase chain reaction. Proc Natl Acad Sci (USA) 1989; 86:9717-9721.

21. Li B, Sehajpal PK, Khanna A, Vlassara H, Cerami A, Stenzel KH, Suthanthiran M. Differential regulation of transforming growth factor beta and interleukin 2 genes in human T cells: Demonstration by usage of novel competitor constructs in the quantitative polymerase chain reaction. J Exp Med 1991; 174:1259-1262.

22. Brillanti S, Garson JA, Tuke PW, Ring C, Briggs M, Masci C, Miglioli M, Barbara L, Tedder RS. Effect of alpha-interferon therapy on hepatitis C viraemia in community acquired non-A, non-B hepatitis. J Med Virol 1991; 34:136-141.

23. Bell J, Ratner L. Specificity of polymerase chain reaction amplification reactions for human immunodeficiency virus type 1 DNA sequences. AIDS Res Hum Retroviruses 1989; 5:87-95.

24. Arrigo SJ, Weitsman S, Rosenblatt JD, Chen IS. Analysis of rev gene function on human immunodeficiency virus type 1 replication in lymphoid cells by using a quantitative polymerase chain reaction method. J Virol 1989; 63:4875-4881.

25. Arrigo SJ, Chen IS. Rev is necessary for translation but not cytoplasmic accumulation of HIV-1 vif, vpr and env/vpu2 RNAs. Genes Dev 1991; 5:808-819.

26. Zack JA, Arrigo SJ, Weitsman SR, Go AS, Haislip A, Chen IS. HIV-1 entry into quiescent primary lymphocytes: Molecular analysis reveals a labile, latent viral structure. Cell 1990; 61:213-222.

27. Zack JA, Haislip AM, Krogstad P, Chen IS. Incompletely reverse transcribed human immunodeficiency virus type 1 genomes in quiescent cells can function as intermediates in the retroviral life cycle. J Virol 1992; 66:1717-1725.

28. Michael NL, Vahey M, Burke DS, Redfield RR. Viral DNA and mRNA expression correlate with stage of human immunodeficiency virus (HIV) type 1 infection in humans: Evidence for viral replication in all stages of HIV disease. J Virol 1992; 66:310-316.

29. D'Addario M, Roulston A, Wainberg MA, Hiscott J: Coordinate enhancement of cytokine gene expression in human immunodeficiency virus type 1-infected promonocytic cells. J Virol 1990; 64:6080-6089.

30. Schnittman SM, Lane HC, Greenhouse J, Justement JS, Baseler M, Fauci AS. Preferential infection of CD4+ memory T-cells by human immunodeficiency virus type 1: Evidence for a role in selective T-cell functional defects observed in infected individuals. Proc Natl Acad Sci 1990; 87:6058-6062.

31. Oka S, Urayama K, Hirabayashi Y, Ohnishi K, Goto H, Mitamura K, Kimura S, Shimada K. Quantitative estimation of human immunodeficiency virus type-1 provirus in CD4+ T lymphocytes using the polymerase chain reaction. Mol Cell Probes 1991; 5:137-142.

32. Pang S, Koyanagi Y, Miles S, Wiley C, Vinters HV, Chen IS. High levels of unintegrated HIV-1 DNA in brain tissue of AIDS dementia patients. Nature 1990; 343:85-89.

33. Rich EA, Chen IS, Zack JA, Leonard ML, O'Brien WA. Increased susceptibility of differentiated mononuclear phagocytes to productive infection with human immunodeficiency virus-1 (HIV-1). J Clin Invest 1992; 89:176-183.

34. Aoki S, Yarchoan R, Thomas RV, Pluda JM, Marczyk K, Broder S, Mitsuya H. Quantitative analysis of HIV-1 proviral DNA in peripheral blood mononuclear cells from patients with AIDS or ARC: Decrease of proviral DNA content following treatment with 2',3'-dideoxyinosine (ddI). AIDS Res Hum Retroviruses 1990; 6:1331-1339.

35. Davis GR, Blumeyer K, DiMichele LJ, Whitfield KM, Chappelle H, Riggs N, Ghosh SS, Kao PM, Fahy E, Kwoh DY, et

al. Detection of human immunodeficiency virus type 1 in AIDS patients using amplification-mediated hybridization analyses: Reproducibility and quantitative limitations. J Infect Dis 1990; 162:13-20.

36. Oka S, Urayama K, Hirabayashi Y, Ohnishi K, Goto H, Mitamura K, Kimura S, Shimada K. Quantitative analysis of human immunodeficiency type-1 DNA in asymptomatic carriers using the polymerase chain reaction. Biochem Biophys Res Comm 1990; 28:1-8.

37. Lee TH, Sunzeri FJ, Tobler LH, Williams BG, Busch MP. Quantitative assessment of HIV-1 DNA load by coamplification of HIV-1 gag and HLA-DQ-alpha genes. AIDS 1991; 5:683-691.

38. Dickover RE, Donovan RM, Goldstein E, Dandekar S, Bush CE, Carlson JR. Quantitation of human immunodeficiency virus DNA by using the polymerase chain reaction. J Clin Microbiol 1990; 28:2130-2133.

39. Jurriaans S, Dekker JT, de Ronde A. HIV-1 viral DNA load in peripheral blood mononuclear cells from seroconverters and long-term infected individuals. AIDS 1992; 6:635-641.

40. Menzo S, Bagnarelli P, Giacca M, Manzin A, Varaldo PE, Clementi M. Absolute quantitation of viremia in human immunodeficiency virus infection by competitive reverse transcription and polymerase chain reaction. J Clin Microbiol 1992; 30:1752-1757.

41. Donovan RM, Dickover RE, Goldstein E, Huth RG, Carlson JR. HIV-1 proviral copy number in blood mononuclear cells form AIDS patients on zidovudine therapy. J Acquir Immune Defic Syndr 1991; 4:766-769.

42. Edlin BR, Weinstein RA, Whaling SM, Ou CY, Connolly PJ, Moore JL, Bitran JD. Zidovudine-interferon-alpha combination therapy in patients with advanced human immunodeficiency type 1 infection: Biphasic response of p24 antigen and quantitative polymerase chain reaction. J Infect Dis 1992; 165:793-798.

43. Dickover RE, Donovan RM, Goldstein E, Cohen SH, Bolton V, Huth RG, Liu GZ, Carlson JR. Decreases in unintegrated HIV DNA are associated with antiretroviral therapy in AIDS patients. J Acquire Immune Defic Syndr 1992; 5:31-36.

44. Ashorn P, Moss B, Weinstein JN, Fitzgerald DJ, Pastan I, Berger EA. Elimination of infectious human immunodeficiency virus form human T-cell cultures by synergistic action of CD4-Pseudomonas exotoxin and reverse transcriptase inhibitors. Proc Natl Acad Sci (USA) 1990; 87:8889-8893.

45. Schnittman SM, Greenhouse JJ, Psallidopoulos MC, Baseler M, Salzman NP, Fauci AS, Lane HC. Increasing viral burden in CD4+ T-cells form patients with human immunodeficiency virus (HIV) infection reflects rapidly progressive immunosuppression and clinical disease. Ann Intern Med 1990; 113:438-443.

46. Ferre F, Marchese A, Duffy PC, Lewis DE, Wallace MR, Beecham HJ, Burnett KG, Jensen FC, Carlo DJ. Quantitation of HIV viral burden by PCR in HIV seropositive Navy personnel representing Walter Reed stages 1 to 6. AIDS Res Hum Retroviruses 1992; 8:269-275.

47. Daar ES, Moudgil T, Meyer RD, Ho DD. Transient high levels of viremia in patients with primary human immunodeficiency virus type 1 infection. N Engl J Med 1991; 324:961-964.

48. Mohler KM, Butler LD. Quantitation of cytokine mRNA levels utilizing the reverse transcriptase polymerase chain reaction following primary antigen-specific sensitization in vivo - I. Verification of linearity, reproducibility and specificity. Mol Immunol 1991; 28:437-447.

49. Kanangat S, Solomon A, Rouse BT. Use of quantitative polymerase chain reaction to quantitate cytokine messenger RNA molecules. Mol Immunol 1992; 29:1229-1236.

50. Platzer C, Richter G, Uberla K, Muller W, Blocker H, Diamantstein T, Blankenstein T. Analysis of cytokine mRNA levels in interleukin-4-transgenic mice by quantitative polymerase chain reaction. Eur J Immunol 1992; 22:1179-1184.

51. Platzer C, Richter G, Uberla K, Hock H, Diamantstein T, Blankenstein T. Interleukin-4-mediated tumor suppression in nude mice involves interferon gamma. Eur J Immunol 1992; 22:1729-1733.

52. Herrmann F, Andreeff M, Gruss HJ, Brach MA, Lubbert M, Mertelsmann R. Interleukin-4 inhibits growth of multiple myelomas by suppressing interleukin-6 expression. Blood 1991; 78:2070-2074.

53. Mullin GF, Lazenby AJ, Harris ML, Bayless TM, James SP. Increased interleukin-2 messenger RNA in the intestinal mucosal lesions of Crohn's disease but not ulcerative colitis. Gastroenterology 1992; 102:1620-1627.

54. Wiles MV, Ruiz P, Imhof BA. Interleukin-7 expression during mouse thymus development. Eur J Immunol 1992; 22:1037-1042.

55. Carre PC, Mortenson RL, King TE Jr, Noble PW, Sable CL, Riches DW. Increased expression of the interleukin-8 gene by alveolar macrophages in idiopathic pulmonary fibrosis. A potential mechanism for the recruitment and activation of neutrophils in lung fibroblasts. J Clin Invest 1991; 88:1802-1810.

56. Enk AH, Katz SI. Identification and induction of keratinocyte-derived IL-10. J Immunol 1992; 149:92-95.

57. Maeda T, Sumida T, Kurasawa K, Tomioka H, Itoh I, Yoshida S, Koike T. T-lymphocyte-receptor repertoire of infiltrating T lymphocytes into NOD mouse pancreas. Diabetes 1991; 40:1580-1585.

58. Sioud M, Kjeldsen-Kragh J, Suleyman S, Vinje O, Natvig JB, Forre O. Limited heterogeneity of T-cell receptor variable region usage in juvenile rheumatoid arthritis synovial T-cells. Eur J Immunol 1992; 22:2413-2418.

59. Roth MS, Antin JH, Bingham EL, Ginsburg D. Use of polymerase chain reaction-detected sequence polymorphisms to document engraftment following allogeneic bone marrow transplantation. Transpl 1990; 49:714-720.

60. Balk SP, Ebert EC, Blumenthal RL, McDermott FV, Wucherpfennig KW, Landau SB, Blumberg RS. Oligoclonal expansion and CD1 recognition by human intestinal intraepithelial lymphocytes. Science 1991; 253:1411-1415.

61. Van Kerckhove C, Russell GJ, Deusch K, Reich K, Bhan AK. Oligoclonality of human intestinal intraepithelial T cells. J Exp Med 1992; 175:57-63.

62. Valles-Ayoub Y, Govan HL, Braun J. Evolving abundance and clonal pattern of human germinal center B cells during childhood. Blood 1990; 76:17-23.

63. Lion T, Izraeli S, Henn T, Gaiger A, Mor W, Gadner H. Monitoring of residual disease in chronic myelogenous leukemia by quantitative polymerase chain reaction. Leukemia 1992; 6:495-499.

64. Dolnick BJ, Zhang ZG, Hines JD, Rustum YM. Quantitation of dihydrofolate reductase and thymidylate synthase mRNAs in vivo and in vitro by polymerase chain reaction. Oncol Res 1992; 4:65-72.

65. Hiroto M, Furukawa Y, Hayashi K. Expression of pS2 gene in human breast cancer cell line MCF-7 is controlled by retinoic acid. Biochem Int 1992; 26:1073-1078.

66. Wiltz O, O'Hara CJ, Steele GD Jr, Mercurio AM. Expression of enzymatically active sucrase-isomaltase is a ubiquitous property of colon adenocarcinomas. Gastroenterology 1991; 100:1266-1278.

67. Wada C, Kasai K, Kameya T, Ohtani H. A general transcription initiation factor, human transcription factor IID, overexpressed in human lung and breast carcinoma and rapidly induced with serum stimulation. Cancer Res 1992; 52:307-313.

68. Carritt B, Kok K, van den Berg A, Osinga J, Pilz A, Hofstra RM, Davis MB, van der Veen AY, Rabbitts PH, Gulati K, et al. A gene from human chromosome region 3p21 with reduced expression in small cell lung cancer. Cancer Res 1992; 52:1536-1541.

69. Delidow RC, White BA, Peluso JJ. Gonadotropin induction of c-fos and c-myc expression and deoxyribonucleic acid synthesis in rat granulosa cells. Endocrinology 1990; 126:2302-2306.

70. Camp TA, Rahal JO, Mayo KE. Cellular localization and hormonal regulation of follicle-stimulating hormone and luteinizing hormone receptor messenger RNAs in the rat ovary. Mol Endocrinol 1991; 5:1405-1417.

71. Park OK, Mayo KE. Transient expression of progesterone receptor messenger RNA in ovarian granulosa cells after the preovulatory luteinizing hormone surge. Mol Endocrinol 1991; 5:967-978.

72. Watson MA, Milbrandt J. Expression of the nerve growth factor-regulated NGFI-A and NGFI-B genes in the developing rat. Development 1990; 110:173-183.

73. Singer-Sam J, Robinson MO, Bellve AR, Simon MI, Riggs AD. Measurement by quantitative PCR of changes in HPRT, PGK-1, PGK2, APRT, MTase and Zfy transcripts during mouse spermatogenesis.

Nucleic Acids Res 1990; 18:1255-1259.

74. Banker DE, Bigler J, Eisenmann RN. The thyroid hormone receptor gene (c-erbA alpha) is expressed in advance of thyroid gland maturation during early embryonic development of Xenopus laevis. Mol Cell Biol 1991; 11:5079-5089.

75. Rupp RA, Weintraub H. Ubiquitous MyoD transcription at the midblastula transition precedes induction-dependent MyoD expression in presumptive mesoderm of X. laevis. Cell 1991; 65:927-937.

76. Talian JC, Zelenka PS. Calpactin I in the differentiating embryonic chicken lens: mRNA levels and protein distribution. Dev Biol 1991; 143:68-77.

77. Mutter GL, Pomponio RJ. Molecular diagnosis of sex chromosome aneuploidy using quantitative PCR. Nucleic Acids Res 1991; 19:4203-4207.

78. Prior TW, Papp AC, Snyder PJ, Highsmith WE Jr, Friedmann KJ, Perry TR, Silverman LM, Mendell JR. Determination of carrier status in Duchenne and Becker muscular dystrophies by quantitative polymerase chain reaction and allele-specific oligonucleotides. Clin Chem 1990; 36:2113-2117.

79. Ioannou P, Christopoulos G, Panayides K, Kleanthous M, Middleton L. Detection of Duchenne and Becker muscular dystrophy carriers by quantitative multiplex polymerase chain reaction analysis. Neurology 1992; 42:1783-1790.

80. Adany R, Iozzo RV. Hypomethylation of the decorin proteoglycan gene in human colon cancer. Biochem J 1991; 276:301-306.

81. Singer-Sam J, Goldstein L, Dai A, Gartler SM, Riggs AD. A potentially critical HpaII site of the X-chromosome-linked PGK1 gene is unmethylated prior to the onset of meiosis of human oogenic cells. Proc Natl Acad Sci (USA) 1992; 89:1413-1417.

82. Ochman H, Gerber AS, Hartl DL. Genetic applications of an inverse polymerase chain reaction. Genetics 1988; 120:621-623.

83. Triglia T, Peterson MG, Kemp DJ. A procedure for in vitro amplification of DNA segments that lie outside the boundaries of known sequences. Nucleic Acids Res 1988; 16:8186.

84. Sels FT, Langer S, Schulz AS, Silver J, Sitbon M, Friedrich RW. Friend murine leukaemia virus is integrated at a common site in most primary spleen tumors of erythroleukaemic animals. Oncogene 1992; 7:643-652.

85. Levy LS, Lobelle-Rich PA. Insertional mutagenesis of flvi-2 in tumors induced by infection with LC-FeLV. a myc-containing strain of feline leukaemia virus. J Virol 1992; 66:2885-2892.

86. Fodde R, Losekoot M, Casula L, Bernini LF. Nucleotide sequence of the Belgian G gamma+ (A gamma delta beta) O-thalassemia deletion breakpoint suggests a common mechanism for a number of such recombination events. Genomics 1990; 8:732-735.

87. Fernandez-Rachubinski F, Rachubinski RA, Blajchman MA. Partial deletion of an antithrombin III allele in a kindred with a type 1 deficiency. Blood 1992; 80:1476-1485.

88. Barr FG, Davis RJ, Eichenfield L, Emanuel BS. Structural analysis of a carcinogen-induced genomic rearrangement event. Proc Natl Acad Sci (USA) 1992; 89:942-946.

89. Bodrug SE, Holden JJ, Ray PN, Worton RG. Molecular analysis of X-autosome translocations in females with Duchenne muscular dystrophy. EMBO J 1991; 10:3931-3939.

90. Huang SJ, Hu YY, Wu CH, Holcenberg J. A simple method for direct cloning cDNA sequence that flanks a region of known sequence from total RNA by applying the inverse polymerase chain reaction. Nucleic Acids Res 1990; 18:1922.

91. Green IR, Sargan DR. Sequence of the cDNA encoding ovine tumor necrosis factor-alpha: Problems with cloning by inverse PCR. Gene 1991; 109:203-210.

92. Nagahashi S, Endoh H, Suzuki Y, Okada N. Characterization of a tandemly repeated DNA sequence family originally derived by retroposition of tRNA(Glu) in the newt. J Mol Biol 1991; 222:391-404.

93. Uematsu Y, Wege H, Straus A, Ott M, Bannwarth W, Lanchbury J, Panayi G, Steinmetz M. The T-cell receptor repertoire in the synovial fluid of a patient with rheumatoid arthritis is polyclonal. Proc Natl Acad Sci (USA) 1991; 88:8534-8538.

94. Berish SA, Mietzner TA, Mayer LW, Genco CA, Holloway BP, Morse SA. Molecular cloning and characterization of the structural gene for the major iron-regulated protein expressed by Neisseria gonorrhoeae. J Exp Med 1990; 171:1535-1546.

95. Rossiter BJ, Fuscoe JC, Muzny DM, Fox M, Caskey CT. The Chinese hamster HPRT gene: Restriction map, sequence analysis, and multiplex PCR deletion screen. Genomics 1991; 9:247-256.

96. Whelan SM, Elmore MJ, Bodsworth NJ, Brehm JK, Atkinson T, Minton NP. Molecular cloning of the Clostridium botulinum structural gene encoding the type B neurotoxin and determination of its entire nucleotide sequence. Appl Environ Microbiol 1992; 58:2345-2354.

97. Erdmann D, Horst G, Dusterhoft A, Kroger M. Stepwise cloning and genetic organization of the seemingly unclonable HgiCII restriction-modification system from Herpetosiphon giganteus strain Hpg9, using PCR technique. Gene 1992; 117:15-22.

98. Mackenzie-Dodds JA, Stamps AC, unpublished observations.

99. Riley J, Butler R, Ogilvie D, Finniear R, Jenner D, Powell S, Anand R, Smith JC and Markham AF. A novel, rapid method for the isolation of terminal sequences from yeast artificial chromosome (YAC) clones. Nucleic Acids Res 1990; 18:2887-2890.

100. Coffey AJ, Roberts RG, Green ED, Cole CG, Butler R, Anand R, Giannelli F, Bentley DR. Construction of a 2.6 Mb contig in yeast artificial chromosomes spanning the human dystrophin gene using an STS-based approach. Genomics 1992; 12:474-484.

101. Roberts RG, Coffey AJ, Bobrow M, Bentley DR. Determination of the exon structure of the distal portion of the dystrophin gene by vectorette PCR. Genomics 1992; 13:942-950.

102. Silverman GA, Jockel JI, Domer PH, Mohr RM, Taillon-Miller P, Korsmeyer SJ. Yeast artificial chromosome cloning of a two-megabase-size contig within chromosomal band 18q21 establishes physical linkage between BCL2 and plasminogen activator inhibitor type-2. Genomics 1991; 9:219-228.

103. Mills KI, Sproul AM, Ogilvie D, Elvin P, Leibowitz D, Burnett AK. Amplification and sequencing of genomic breakpoints located within the M-bcr region by Vectorette-mediated polymerase chain reaction. Leukemia 1992; 6:481-483.

104. Espelund M, Jakobsen KS. Cloning and direct sequencing of plant promoters using primer-adapter mediated PCR on DNA coupled to a magnetic solid phase. Biotechniques 1992; 13:74-81.

105. Compton T. Degenerate primers for DNA amplification. In: Innis MA, Gelfand DH, Sninsky JJ and White TJ, eds. PCR Protocols. San Diego, California: Academic Press, Inc, 1990: 39-45.

106. Lee CC, Wu X, Gibbs RA, Cook RG, Muzny DM, Caskey CT. Generation of cDNA probes directed by amino acid sequence: Cloning of urate oxidase. Science 1988; 239:1288-1291.

107. Strub K, Walter P. Isolation of a cDNA clone of the 14-kDa subunit of the signal recognition particle by cross-hybridization of differently primed polymerase chain reactions. Proc Natl Acad Sci (USA) 1989; 86:9747-9751.

108. Gonzalez GA, Yamamoto KK, Fischer WH, Karr D, Manzel P, Biggs W, Vale WW, Montminy MR. A cluster of phosphorylation sites on the cyclic AMP-regulated nuclear factor CREB predicted by its sequence. Nature 1989; 337:749-752.

109. Kopin AS, Wheeler MB, Leiter AB. Secretin: Structure of the precursor and tissue distribution of the mRNA. Proc Natl Acad Sci (USA) 1990; 87:2299-2303.

110. Larzul D, Guigue F, Sninsky JJ, Mack DH, Brechot C, Guesdon J-L. Detection of hepatitis B virus sequences in serum by using in vitro enzymatic amplification. J Virol Methods 1988; 20:227-237.

111. Girgis SI, Alevizaki M, Denny P, Ferrier GJ, Legon S. Generation of DNA probes for peptides with highly degenerate codons using mixed primer PCR. Nucleic Acids Res 1988; 16:10371.

112. Knoth K, Roberds S, Poteet C, Tamkun M. Highly degenerate, inosine-containing primers specifically amplify rare cDNA using the polymerase chain reaction. Nucleic Acids Res 1988; 16:10932.

113. Patil RV, Dekker EE. PCR amplification of an Escherichia coli gene using mixed primers containing deoxyinosone in degenerate amino acid codons. Nucleic Acids Res 1990; 18:3080.

114. Heller RA, Song K, Freire-Moar J. Rapid screening of libraries with radiolabeled DNA sequences generated by PCR using highly degenerate oligonucleotide mixtures. Biotechniques 1992; 12:30,32,34,35.

115. Drayna D, Davies KE, Hartley DA, Mandel JL, Camerino G, Williamson R, White RL. Genetic mapping of the human X chromosome by using restriction fragment length polymorphisms. Proc Natl Acad Sci (USA) 1984; 81:2836-2839.

116. Wilks AF. Two putative protein tyrosine kinases identified by application of the polymerase chain reaction. Proc Natl Acad Sci (USA) 1989; 86:1603-1607.

117. Wilks AF, Kurban RR, Hovens CM, Ralph SJ. The application of the polymerase chain reaction to cloning members of the protein tyrosine kinase family. Gene 1989; 85:67-74.

118. LeBoeuf RD, Galin FS, Hollinger SK, Peiper SC, Blalock JE. Cloning and sequencing of immunoglobulin variable region genes using degenerate oligodeoxyribonucleotides and polymerase chain reaction. Gene 1989; 82:371-377.

119. Kamb A, Weir M, Rudy B, Varmus H, Kenyon C. Identification of genes from pattern formation, tyrosine kinase, and potassium channel families by DNA amplification. Proc Natl Acad Sci (USA) 1989; 86:4372-4376.

120. Chernoff J, Schievella AR, Jost CA, Erikson RL, Neel BG. Cloning of a cDNA for a major human protein tyrosine phosphatase. Proc Natl Acad Sci (USA) 1990; 87:2735-2739.

121. Schaap D, de Widt J, van der Wal J, Vandekerckhove J, van Damme J, Gussow D, Ploegh HL, van Blitterswijk WJ, van der Bend RL. Purification, cDNA cloning and expression of human diacylglycerol kinase. FEBS Lett 1990; 275:151-158.

122. Loetscher H, Pan YC, Lahm HW, Gentz R, Brockhaus M, Tabuchi H, Lesslauer W. Molecular cloning and expression of the human 55 kd tumor necrosis factor receptor. Cell 1990; 61:351-359.

123. Wank SA, Harkins R, Jensen RT, Shapira H, de Weerth A, Slattery T. Purification, molecular cloning, and functional expression of the cholecystokinin receptor from rat pancreas. Proc Natl Acad Sci (USA) 1992; 89:3125-3129.

124. Chen ZJ, Niles EG, Pickart CM. Isolation of a cDNA encoding a mammalian multiubiquitinating enzyme (E225K) and overexpression of the functional enzyme in Escherichia coli. J Biol Chem 1991; 266:15698-15704.

125. Bischoff J, Moremen K, Lodish HF. Isolation, characterization, and expression of cDNA encoding a rat liver endoplasmic reticulum alpha-mannosidase. J Biol Chem 1990; 265:17110-17117.

126. Keller PM, Schultz LD, Condra C, Karczewski J, Conolly TM. An inhibitor of collagen-stimulated platelet activation from the salivary glands of the Haementeria officinalis leech. II. Cloning of the cDNA and expression.J Biol Chem 1992; 267:6899-6904.

127. Morgenstern JP, Griffith IJ, Brauer AW, Rogers BL, Bond JF, Chapman MD, Kuo MC. Amino acid sequence of Fel dI, the major allergen of the domestic cat: Protein sequence analysis and cDNA cloning. Proc Natl Acad Sci (USA) 1991; 88:9690-9694.

128. Briggs JA, Burrus GR, Stickney BD, Briggs RC. Cloning and expression of the human myeloid cell nuclear differentiation antigen: Regulation by interferon alpha. J Cell Biochem 1992; 49:82-92.

129. Sugiura N, Hagiwara H, Hirose S. Molecular cloning of porcine soluble angiotensin-binding protein. J Biol Chem 1992; 267:18067-18072.

130. Price HM, Ryan RO, Haunerland NH. Primary structure of locust flight muscle fatty acid binding protein. Arch Biochem Biophys 1992; 297:285-290.

131. Huang GC, Page MJ, Roberts AJ, Malik AN, Spence H, McGregor AM, Banga JP. Molecular cloning of human thyrotropin receptor cDNA fragment. Use of highly degenerate, inosine containing primers derived from aligned amino acid sequences of a homologous family of glycoprotein hormone receptors. FEBS Lett 1990; 264:193-197.

132. Gantz I, Schaffer M, DelValle J, Logsdon C, Campbell V, Uhler M, Yamada T. Molecular cloning of a gene encoding the histamine H2 receptor. Proc Natl Acad Sci (USA) 1991; 88:429-433.

133. Monsma FJ Jr, Mahan LC, McVittie LD, Gerfen CR, Sibley DR. Molecular cloning and expression of a D1 dopamine receptor linked to adenylyl cyclase activation. Proc Natl Acad Sci (USA) 1990; 87:6723-6727.

134. Yoshimura M, Cooper DM. Cloning and expression of a Ca(2+)-inhibitable adenylyl cyclase from NCB-20 cells. Proc Natl Acad Sci (USA) 1992; 89:6716-6720.

135. Mahan LC, McVittie LD, Smyk-Randall EM, Nakata H, Monsma FJ Jr, Gerfen CR, Sibley Jr. Cloning and expression of an A1 adenosine receptor from rat brain. Mol Pharmacol 1991; 40:1-7.

136. Bruno JF, Whittaker J, Song JF, Berelowitz M. Molecular cloning of a cDNA encoding a human alpha 1A adrenergic receptor. Biochem Biophys Res Commun 1991; 179:1485-1490.

137. Gerard CM, Mollereau C, Vassart G, Parmentier M. Molecular cloning of a human cannabinoid receptor which is also expressed in testis. Biochem J 1991; 279:129-134.

138. Rimland J, Xin W, Sweetnam P, Saijoh K, Nestler EJ, Duman RS. Sequence and expression of a neuropeptide Y receptor cDNA. Mol Pharmacol 1991; 40:869-875.

139. McAllister G, Charlesworth A, Snodin C, Beer MS, Noble AJ, Middlemiss DN, Iversen LL, Whiting P. Molecular cloning of a serotonin receptor from human brain (5HT1E): A fifth 5HT1-like subtype. Proc Natl Acad Sci (USA) 1992; 89:5517-5521.

140. Libert F, Parmentier M, Lefort A, Dinsart C, Van Sande J, Maenhaut C, Simons MJ, Dumont JE, Vassart G. Selective amplification and cloning of four new members of the G protein-coupled receptor family. Science 1989; 244:569-572.

141. Eidne KA, Zabavnik J, Peters T, Yoshida S, Anderson L, Taylor PL. Cloning, sequencing and tissue distribution of a candidate G protein-coupled receptor from rat pituitary gland. FEBS Lett 1991; 292:243-248.

142. Wilks AF, Harpur AG, Kurban RR, Ralph SJ, Zurcher G, Ziemiecki A. Two novel protein tyrosine kinases, each with a second phosphotransferase-related catalytic domain, define a new class of protein kinase. Mol Cell Biol 1991; 11:2057-2065.

143. Howard OM, Dean M, Young H, Ramsburg M, Turpin JA, Michiel DF, Kelvin DJ, Lee L, Farrar WL. Characterization of a class 3 tyrosine kinase. Oncogene 1992; 7:895-900.

144. Terman BI, Carrion ME, Kovacs E, Rasmussen BA, Eddy RL, Shows TB. Identification of a new endothelial cell growth factor receptor tyrosine kinase. Oncogene 1991; 6:1677-1683.

145. Nishi M, Ohagi S, Steiner DF. Novel putative protein tyrosine phosphatases identified by the polymerase chain reaction. FEBS Lett 1990; 271:178-180.

146. Ottilie S, Chernoff J, Hannig G, Hoffman CS, Erikson RL. A fission yeast gene encoding a protein with features of protein tyrosine phosphatases. Proc Natl Acad Sci (USA) 1991; 88:3455-3459.

147. Blakely RD, Berson HE, Fremeau RT Jr, Caron MG, Peek MM, Prince HK, Bradley CC. Cloning and expression of a functional serotonin transporter from rat brain. Nature 1991; 354:66-70.

148. Fremeau RT Jr, Caron MG, Blakely RD. Molecular cloning and expression of a high affinity L-proline transporter expressed in putative glutamatergic pathways of rat brain. Neuron 1992; 8:915-926.

149. Mayser W, Betz H, Schloss P. Isolation of cDNAs encoding a novel member of the neurotransmitter transporter gene family. FEBS Lett 1991; 295:203-206.

150. Repaske DR, Swinnen JV, Jin SL, Van Wyk JJ, Conti M. A polymerase chain reaction strategy to identify and clone cyclic nucleotide phosphodiesterase cDNAs. Molecular cloning of the cDNA encoding the 63-kDa calmodulin-dependent phosphodiesterase. J Biol Chem 1992; 267:18683-18688.

151. Chen T, Bunting M, Karim FD, Thummel CS. Isolation and characterization of five Drosophila genes that encode an ets-related DNA binding domain. Dev Biol 1992; 151:176-191.

152. Endow SA, Hatsumi M. A multimember kinesin gene family in Drosophila. Proc Natl Acad Sci (USA) 1991; 88:4424-4427.

153. Mackem S, Mahon KA. Ghox 4.7: A chick homeobox gene expressed primarily in limb

buds with limb-type differences in expression. Development 1991; 112:791-806.

154. Stewart RJ, Pesavento PA, Woerpel DN, Goldstein LS. Identification and partial characterization of six members of the kinesin superfamily in Drosophila. Proc Natl Acad Sci (USA) 1991; 88:8470-8474.

155. Le Moine C, Young WS. RHS2, aPOU domain-containing gene, and its expression in developing and adult rat. Proc Natl Acad Sci (USA) 1992; 89:3285-3289.

156. Frohman MA, Downs TR, Chomczynski P, Frohman LA. Cloning and characterization of mouse growth hormone-releasing hormone (GRH) complementary DNA: Increased GRH messenger RNA levels in the growth hormone-deficient lit/lit mouse. Mol Endocrinol 1989; 3:1529-1536.

157. Ernfors P, Ibanez CF, Ebendal T, Olson L, Persson H. Molecular cloning and neurotrophic activities of a protein with structural similarities to nerve growth factor: Developmental and topographical expression in the brain. Proc Natl Acad Sci (USA) 1990; 87:5454-5458.

158. Zhao D, Yang J, Jones KE, Gerald C, Suzuki Y, Hogan PG, Chin WW, Tashjian AH Jr. Molecular cloning of a complementary deoxyribonucleic acid encoding the thyrotropin-releasing hormone receptor and regulation of its messenger ribonucleic acid in rat GH cells. Endocrinol 1992; 130:3529-3526.

159. Vallins WJ, Brand NJ, Dabhade N, Butler-Browne G, Yacoub MH, Barton PJ. Molecular cloning of human cardiac troponin I using polymerase chain reaction. FEBS Lett 1990; 270:57-61.

160. Sun W, Ferrer-Montiel AV, Schinder AF, McPherson JP, Evans GA, Montal M. Molecular cloning, chromosomal mapping, amd functional expression of human brain glutamate receptors. Proc Natl Acad Sci (USA) 1992; 89:1443-1447.

161. Tseng-Crank JC, Tseng GN, Schwartz A, Tanouye MA. Molecular cloning and functional expression of a potassium channel cDNA isolated from a rat cardiac library. FEBS Lett 1990; 268:63-68.

162. Ruben SM, Dillon PJ, Schreck R, Henkel T, Chen CH, Maher M, Baeuerle PA, Rosen CA. Isolation of a *rel*-related human cDNA that potentially encodes the 65-kD subunit of NF-kappa B. Science 1991; 251:1490-1493.

163. Ghiara JB, Richardson HE, Sugimoto K, Henze M, Lew DJ, Wittenberg C, Reed SI. A cyclin B homolog in S. cerevisiae: Chronic activation of the Cdc28 protein kinase by cyclin prevents exit from mitosis. Cell 1991; 65:163-174.

164. Wei W, Mosteller RD, Sanyal P, Gonzales E, McKinney D, Dasgupta C, Li P, Liu BX, Broek D. Identification of a mammalian gene structurally and functionally related to the CDC25 gene of Saccharomyces cerevisiae. Proc Natl Acad Sci (USA) 1992; 89:7100-7104.

165. Posas F, Casamayor A, Morral N, Arino J. Molecular cloning and analysis of a yeast protein phosphatase with an unusual amino-terminal region. J Biol Chem 1992; 267:11734-11740.

166. Hata S, Tsukamoto T, Osumi T, Hashimoto J, Suzuka I. Analysis of carrot genes for proliferating cell nuclear antigen homologs with the aid of the polymerase chain reaction. Biochem Biophys Res Commun 1992; 184:576-581.

167. Feiler HS, Jacobs TW. Cloning of the pea cdc2 homologue by efficient immunological screening of PCR products. Plant Mol Biol 1991; 17:321-333.

168. Engel JN, Pollack J, Malik F, Ganem D. Cloning and characterization of RNA polymerase core subunits of Chlamydia trachomatis by using the polymerase chain reaction. J Bacteriol 1990; 172:5732-5741.

169. Gooze L, Kim K, Petersen C, Gut J, Nelson RG. Amplification of a Cryptosporidium parvum gene fragment encoding thymidylate synthase. J Protozool 1991; 38:56S-58S.

170. Sakanari JA, Staunton CE, Eakin AE, Craik CS, McKerrow JH. Serine proteases from nematode and protozoan parasites: Isolation of sequence homologs using generic molecular probes. Proc Natl Acad Sci (USA) 1989; 86:4863-4867.

171. Nunberg JH, Wright DK, Cole GE, Petrovskis EA, Post LE, Compton T, Gilbert JH. Identification of the thymidine kinase gene of feline herpesvirus: Use of degenerate oligonucleotides in the polymerase chain reaction to isolate herpesvirus gene homologs. J Virol 1989; 63:3240-3249.

172. Keeler CL Jr, Kingsley DH, Burton CR. Identification of the thymidine kinase gene of infectious laryngotracheitis virus. Avian Dis 1991; 35:920-929.

173. Teo IA, Griffin BE, Jones MD. Characterization of the DNA polymerase gene of human herpesvirus 6. J Virol 1991; 65:4670-4680.

174. Donehower LA, Bohannon RC, Ford RJ, Gibbs RA. The use of primers from highly conserved pol regions to identify uncharacterized retroviruses by the polymerase chain reaction. J Virol Methods 1990; 28:33-46.

175. Snijders PJ, Meijer CJ, Walboomers JM. Degenerate primers based on highly conserved regions of amino acid sequence in papillomaviruses can be used in a generalized polymerase chain reaction to detect productive human papillomavirus infection. J Gen Virol 1991; 72:2781-2786.

176. Williamson AL, Jaskiesicz K, Gunning A. The detection of human papillomavirus in oesophageal lesions. Anticancer Res 1991; 11:263-265.

177. Williamson AL, Rybicki EP. Detection of human papillomaviruses by polymerase chain reaction amplification with degenerate nested primers. J Med Virol 1991; 33:165-171.

178. Evander M, Edlund K, Boden E, Gustafsson A, Jonsson M, Karlsson R, Rylander E, Wadell G. Comparison of a one-step and a two-step polymerase chain reaction with degenerate general primers in a population-based study of human papillomavirus infection in young Swedish women. J Clin Microbiol 1992; 30:987-992.

179. Brachman DG, Graves D, Vokes E, Beckett M, Haraf D, Montag A, Dunphy E, Mick R, Yandell D, Weichelsbaum RR. Occurrence of p53 gene deletions and human papillomavirus infection in human head and neck cancer. Cancer Res 1992; 52:4832-4836.

180. Henchal EA, Polo SL, Vorndam V, Yaemsiri C, Innis BL, Hoke CH. Sensitivity and specificity of a universal primer set for the rapid diagnosis of dengue virus infections by polymerase chain reaction and nucleic acid hybridization. Am J Trop Med Hyg 1991; 45:418-428.

181. Kaltenboeck B, Kousoulas KG, Storz J. Two-step polymerase chain reactions and restriction endonuclease analyses detect and differentiate ompA DNA of Chlamydia spp. J Clin Microbiol 1992; 30:1098-1104.

182. Meredith SE, Lando G, Gbakima AA, Zimmerman PA, Unnasch TR. Onchocerca volvulus: Application of the polymerase chain reaction to identification and strain differentiation of the parasite. Exp Parasitol 1991; 73:335-344.

183. Saiki RK, Gelfand DH, Stoffel S, Scharf S, Higuchi R, Horn GT, Mullis K, Erlich HA. Primer-directed enzymatic amplification of DNA with a thermostable DNA polymerase. Science 1988; 239:487-491.

184. Sanger F, Nicklen S, Coulson AR. DNA sequencing with chain-terminating inhibitors. Proc Natl Acad Sci (USA) 1977; 74:5463-5468.

185. Wrishik LA, Higuchi RG, Stoneking M, Erlich HA, Arnheim N, Wilson AC. Length mutations in human mitochondrial DNA: direct sequencing of enzymatically amplified DNA. Nucleic Acids Res 1987; 15:529-542.

186. Newton CR, Kalsheker N, Graham A, Powell S, Gammack A, Riley J, Markham AJ. Diagnosis of alpha$_1$-antitrypsin deficiency by enzymatic amplification of human genomic DNA, direct sequencing of polymerase chain reaction products. Nucleic Acids Res 1988; 16:8233-8243.

187. Higuchi R, von Beroldingen CH, Sensabaugh GF, Erlich HA. DNA typing from single hairs. Nature 1988; 332:543-546.

188. Innis MA, Myambo KB, Gelfand DH, Brow MAD. DNA sequencing with Thermus aquaticus DNA polymerase and direct sequencing of polymerase chain reaction-amplified DNA. Proc Natl Acad Sci (USA) 1988; 85:9436-9440.

189. Gyllensten UB, Erlich HA. Generation of single-stranded DNA by the polymerase chain reaction and its application to direct sequencing of the HLA-DQA locus. Proc Natl Acad Sci (USA) 1988; 85:7652-7656.

190. McCabe P. Production of single-stranded DNA by asymmetric PCR. In: Innis MA, Gelfand DH, Sninsky JJ, White TJ, eds: PCR Protocols. San Diego, California: Academic Press 1990: 76-83.

191. Bartek J, Iggo R, Gannon J, Lane DP. Genetic and immunochemical analysis of mutant p53 in human breast cancer cell

lines. Oncogene 1990; 5:893-899.

192. Hultman T, Stahl S, Hornes E, Uhlen M. Direct solid phase sequencing of genomic and plasmid DNA using magnetic beads as solid support. Nucleic Acids Res 1989; 17:4937-4946.

193. Gusterson BA, Anbazhagan R, Warren W, Midgely C, Lane DP, O'Hare M, Stamps A, Carter R, Jayatilake H. Expression of p53 in premalignant and malignant squamous epithelium. Oncogene 1991; 6:1785-1789.

194. Warren W, Eeles RA, Ponder BAJ, Easton DF, Averill D, Ponder MA, Anderson K, Evans AM, DeMars R, Love R, Dundas S, Stratton MR, Trowbridge P, Cooper CS, Peto J. No evidence for germline mutations in exons 5-9 of the p53 gene in 25 breast cancer families. Oncogene 1992; 7:1043-1046.

195. Lu X, Park SH, Thompson TC, Lane DP. Ras-induced hyperplasia occurs with mutation of p53, but activated ras and myc together can induce carcinoma without p53 mutation. Cell 1992; 70:153-161.

196. Ruan CC, Fuller CW. Using T7 gene 6 exonuclease to prepare single-stranded templates for sequencing. Editorial Comments 1991; 18:1-8. United States Biochemical Corp.

197. Kerr C, Sadowski PD. Gene 6 exonuclease of bacteriophage T7; purification and properties of the enzyme. J Biol Chem 1972; 247:305-310.

198. Engelke DR, Hoener PA, Collins FS. Direct sequencing of enzymatically amplified human genomic DNA. Proc Natl Acad Sci (USA) 1988; 85:544-548.

199. Murray V. Improved double-stranded DNA sequencing using the linear polymerase chain reaction. Nucleic Acids Res 1989; 17:8889.

200. Krishnan BR, Blakesley RW, Berg DE. Linear amplification DNA sequencing directly from single phage plaques and bacterial colonies. Nucleic Acids Res 1991; 19:1153.

201. Krishnan BR, Kersulyte D, Brikun I, Berg CM, Berg DE. Direct and crossover PCR amplification to facilitate Tn5supF-based sequencing of lambda phage clones. Nucleic Acids Res 1991; 19:6177-6182.

202. Simpson D, Crosby RM, Skopek TR. A method for specific cloning and sequencing of human hprt cDNA for mutation analy-

sis. Biochem Biophys Res Commun 1988; 151:487-492.

203. Vrieling H, Simons JW, van Zeeland AA. Nucleotide sequence determination of point mutations at the mouse HPRT locus using in vitro amplification of HPRT mRNA sequences. Mutat Res 1988; 198:107-113.

204. Scharf SJ, Long CM, Erlich HA. Sequence analysis of the HLA-DR beta and HLA-DQ beta loci from three Pemphigus vulgaris patients. Hum Immunol 1988; 22:61-69.

205. Newton CR, Kalsheker N, Graham A, Powell S, Gammack A, Riley J, Markham AF. Diagnosis of alpha 1-antitrypsin deficiency by enzymatic amplification of human genomic DNA and direct sequencing of polymerase chain reaction products. Nucleic Acids Res 1988; 16:8233-8243.

206. Labhard ME, Wirtz MK, Pope FM, Nicholls AC, Hollister DW. A cysteine for glycine substitution at position 1017 in an alpha 1(I) chain of type I collagen in a patient with mild dominantly inherited osteogenesis imperfecta. Mol Biol Med 1988; 5:197-207.

207. Cowman Af, Morry MJ, Biggs BA, Cross GA, Foote SJ. Amino acid changes linked to pyrimethamine resistance in the dihydrofolate reductase-thymidylate synthase gene of Plasmodium falciparum. Proc Natl Acad Sci (USA) 1988; 85:9109-9113.

208. Clark JM. Novel non-templated nucleotide addition reactions catalyzed by procaryotic and eucaryotic DNA polymerases. Nucleic Acids Res 1988; 20:9677-9686.

209. Mole SE, Iggo RD, Lane DP. Using the polymerase chain reaction to modify expression plasmids for epitope mapping. Nucleic Acids Res 1989; 17:3319.

210. Hemsley A, Arnheim N, Toney MD, Cortopassi G, Galas DJ. A simple method for site-directed mutagenesis using the polymerase chain reaction. Nucleic Acids Res 1989; 17:6545-6551.

211. Marchuk D, Drumm M, Saulino A, Collins FS. Construction of T-vectors, a rapid and general system for direct cloning of unmodified PCR products. Nucleic Acids Res 1991; 19:1154.

212. Holton TA, Graham MW. A simple and efficient method for direct cloning of PCR products using ddT-tailed vectors. Nucleic Acids Res 1991; 19:1156.

213. Kozak M. The scanning model for translation: An update. J Cell Biol 1989; 108:229-241.

214. Goodenow M, Huet T, Saurin W, Kwok S, Sninsky J, Wain-Hobson S. HIV1 isolates are rapidly evolving quasispecies: Evidence for viral mixtures and preferred nucleotide substitutions. JAIDS 1989; 2:344-352.

215. Fucharoen S, Fucharoen G, Fucharoen P, Fukumaki Y. A novel ochre mutation in the β-thalassemia gene of a Thai. J Biol Chem 1989; 264:7780-7783.

216. Mor O, Grossman Z, Jakobovitz O, Brok-Simoni F, Rechavi G. Artifactual p53 point mutations: Possible effect of gene secondary structure on PCR and direct sequence analysis. Lancet 1992; 340:1236.

217. Lundberg KS, Shoemaker DD, Adams MW, Short SM, Sarge JA, Mathur EJ. High fidelity amplification using a thermostable DNA polymerase isolated from Pyrococcus furiosus. Gene 1991; 108:1-6.

INDEX